THE WEARABLE TECHNOLOGY HANDBOOK

Second (Metaverse) Edition

Haider Raad, Ph.D.

Ohio Publishing and Academic Services, USA
Our distinguished editorial team thrives to produce and publish superior quality books.
Visit our website for the current list of publications
www.ohiopas.org

Author: Haider Raad, Ph.D.
Department of Physics, Engineering Phyiscs Program
Xavier Wearable Electronics Research Center (XWERC)
Xavier University, Cincinnati, Ohio, USA

Published by Ohio Publishing and Academic Services, USA
Copyright © 2022 Ohio Publishing and Academic Services
www.ohiopas.org
info@ohiopas.org

All rights reserved. No part of this book may be reproduced, copied (electronically or mechanically), or videotaped without a written permission of the publisher.

ISBN: 978-1-7372334-8-0

Disclaimer
The Publisher and the Author hold no liability for incidental or consequential injuries or damages caused by the information contained in this publication.

ABOUT THE AUTHOR

 Dr. Haider Raad currently serves as the director of the Engineering Physics program and the Wearable Electronics Research Center (XWERC) at Xavier University in Ohio, USA. He was previously affiliated with California State University & the University of Arkansas, Little Rock between 2008 and 2015. Haider received the Ph.D. and M.S. degrees in Systems Engineering, specializing in RF Telecommunication and Wireless Systems from the University of Arkansas at Little Rock (UALR), and the M.S. degree in Electrical and Computer Engineering from New York Institute of Technology (NYIT).

Professor Raad teaches several courses such as Electronic Circuits, Microprocessors and Digital Systems, Communication Systems, Antenna Engineering, and Control Theory. He has given over 50 lectures at universities around the world and is a frequent speaker at international conferences. Professor Raad is also connected to the industry through his engineering consulting firm.

Dr. Raad has published seven books in the fields of Wearable Technology, Telemedicine, and Wireless Systems. He has also published over a hundred peer reviewed journal and conference papers on research fields of his interest which include: Flexible and Wearable Wireless Systems, Telemedicine and Wireless Body Area Networks, IoT, Metamaterials, and Biomedical Electronics. He is also the recipient of the Province Ignatian Educator of Distinction Award (2020), the Outstanding Teaching Award (2019), the 19th International Wearable Technology Conference Best Paper Award (2017), the E-Telemed Conference

Best Paper Award (2016), Sonoma State University's Research Fellowship Award (2015), and AAMI/TEAMS Academic Excellence Award (2012).

Haider is also a professional musician who goes by the stage name "Doctor X"! He enjoys composing world, rock, and smooth jazz music. He also enjoys travelling, street photography, and other artistic activities.

To my family with love

Preface

Everything will be connected. This is one of the rules that will govern the future. And contrary to popular belief, the impact of Wearable Technology will be much greater than a smart watch or a fitness tracker. Connecting everything will dramatically reshape our world in ways we can barely imagine.

Locating a wandering Alzheimer's patient by sensors embedded within lighting poles in a smart city, or detecting if the driver is having a heart attack by analyzing vital signs and facial expressions by a system integrated within a vehicle's dashboard, are just a couple of scenarios these technologies will be capable of doing. We will also witness the fantasy of fully automated smart cities and driverless vehicles work in coordination with one another fairly soon. Further, an extremely hot topic of conversation today is "the Metaverse", which is a network of 3D digital worlds aimed at creating rich user interaction and immersive experiences that mimic the real world. What makes this subject serious is that giants like Meta (formerly: Facebook), and Microsoft are both staking claims. Obviously, without Wearable Technology to bridge between the physical and virtual worlds, the Metaverse will be nothing but an unattainable fantasy.

Today, Wearable Technology is recognized as one of the fastest growing technologies and hottest research topics in academia and research and development centers. Wearable devices, which are characterized by being light-weight, energy efficient, ergonomic, and potentially reconfigurable are expected to substantially expand the applications of modern consumer electronics.

With around 18 billion devices connected to the Internet as of 2018, recent market studies speculate that this number could be in the range of 100 billion "things" connected in the next few years. Another report indicated that the global shipments of wearables reached 49.6 million units in 2019, 55.2% up from 2018, with smart watches and wristbands continuing to dominate the wearables landscape, accounting for 63.2% of all devices shipped in this year. It is anticipated that the global wearables market share will exceed $116 billion in 2025.

Having worked in this field for almost 14 years in both academic and business capacities, I felt the need to compile a complete resource that tackles all aspects of this vital technology. Also, due to the fast-paced technological advancement in this area, I felt that it's time to release the second edition of this book, with updated material and an entire chapter dedicated to discuss the Metaverse.

The aim of this book, as the title suggests, is to provide a comprehensive guide to various components and applications of wearable technology, in addition to its social, and psychological impacts. Moreover, privacy, security, and health concerns will also be covered in this book.

The intended audience of this book includes, but not limited to, scientists in the Research and Development field, university professors, practicing technologists, in addition to all the enthusiasts interested in this fascinating technology. Moreover, the book serves as an extensive resource for both undergraduate and graduate students working on topics related to Wearable Technology and the Metaverse.

Haider Raad, Ph.D.
Xavier University, USA
May 2022

Table of Contents

Chapter 1: Introduction and Historical Background 1

1. Introduction 1
    1.1 What Exactly is a Wearable Device 2
    1.2 Characteristics of Wearable Devices 4
    1.3 IoT vs. Wearables 6
2. A Brief History of Wearable Technology 7
3. The Wearables We Know Today 12
4. Challenges 18
    4.1 Security 19
    4.2 Privacy 20
    4.3 Standards and Regulations 21
    4.4 Energy and Power Issues 22
    4.5 Connectivity 22
Conclusion 23
References 24

Chapter 2: Applications of Wearable Technology 27

1. Introduction 27
2. Applications 27
    2.1 Biomedicine and Healthcare 27
        2.1.1 Adhesives 33
    2.2 Fitness and Wellbeing 34
    2.3 Sports 37
    2.4 Entertainment 40
    2.5 Gaming 42
    2.6 Pets 43
    2.7 Military and Public Safety 44
    2.8 Civil Defense 45
    2.9 Travel and Tourism 46
    2.10 Aerospace 46
    2.11 Business and Industry 47
    2.12 Education 49
    2.13 Fashion 50

2.14 Novel and Unusual Applications................................54
References 55

Chapter 3: Components and Technologies 63

Introduction 63
1. Hardware 65
   1.1 Sensors 65
        1.1.1 Sensor Properties 67
        1.1.2 MEMS Sensors 69
        1.1.3 Wireless Sensors..75
        1.1.4 Multisensor Modules....................................75
        1.1.5 Signal Conditioning for Sensors...................77
   1.2 Actuators 77
   1.3 Microprocessors and Microcontrollers 78
   1.3.1 Selecting the Right Processing Unit 84
   1.4 Wireless Connctivity Unit 86
   1.5 Battery Technology 89
        1.5.1 Power Management Circuits 91
   1.6 Displays and other UI Elecments 92
   1.7 Microphones and Speakers 93
2. Architectures, Software, and Communication Tech. ...93
   2.1 IoT Architectures 94
   2.2 Wearable Device Architecture..............................95
   2.3 Operating Systems 96
   2.4 Communication Procols and Technologies 97
        2.4.1 Short Range...98
        2.4.1.1 Bluetooth..98
        2.4.1.2 NFC and RFID 99
        2.4.1.3 Z-Wave...99
        2.4.2 Medium Range...100
        2.4.2.1 Wi-Fi..100
        2.4.2.2 ZigBee 100
        2.4.3 Long Range..101
        2.4.3.1 LPWAN and LORA................................101
        2.4.3.2 Sigfox...101
        2.4.3.3 Cellular Technologies 101
   2.5 Cloud 102
        2.5.1 Why Cloud? 103

  2.5.2 Platforms 104
   2.5.2.1 Criteria for Choosing a Platform 106
  2.6 Data Analytics and Machine Learning 107
   2.6.1 Artificial Intelligence 108
   2.6.2 Machine Learning 109
   2.6.3 Data Mining 109
  2.7 Other Software Technologies 110
   2.7.1 Virtual and Augmented Reality 110
   2.7.2 Voice Recognition 110
Conclusion 111
References 112

Chapter Four: Product Development and Design Consideration 117

1. Introduction 117
2. Product Development Process 118
  2.1 Ideation and Research 118
  2.2 Requirements/Specifications...................................118
  2.3 Engineering Analysis 118
   2.3.1 Hardware Design 119
   2.3.2 Software Development................................120
   2.3.3 Mechanical Design.....................................120
  2.4 Prototyping..123
  2.5 Testing and Validation...123
   2.5.1 Review and Design Verification...................124
   2.5.2 Unit Testing..124
   2.5.3 Integration Testing 125
   2.5.4 Certification and Documentation 125
   2.5.5 Production Review 126
  2.4 Production..126
3. Wearable Product Requirements 126
  3.1 Form Factor..128
  3.2 Power Requirements..128
   3.2.1 Power Budget 129
  3.3 Wireless Connectivity Requirements 131
   3.3.1 RF Design & Antenna Matching 132
   3.3.2 Link Budget 133
  3.4 Cost Requirements...136

4. Design Considerations 136
4.1 Operational Factors 136
4.2 Durability and Longevity 136
4.3 Reliability...137
4.4 Usability and User Interface 138
4.5 Aesthetics 138
4.6 Compatibility...139
4.7 Comfort and Ergonomic Factors 139
4.8 Safety Factors 140
4.9 Washing factors (Wash-ability)..........................140
4.10 Maintenance Factors 140
4.11 Packaging and Material Requirements................141
4.12 Security Factors..141
4.13 Technology Obsolescence..............................142
Conclusion...142
References...144

Chapter Five: Security Issues and Privacy Concerns 147

1. Introduction 147
2. Security and Privacy Issues in Wearable Technology................149
    2.1 Privacy and Security Concerns in Digital Technologies
....151
    2.2 Threats and Attacks 153
    2.3 Threat Modeling..154
    2.4 Common Attacks 156
    2.5 Privacy Issues..159
    2.6 Potential Solutions 160
References 163

Chapter Six: Psychological and Social Impacts 165

1. Introduction 165
2. The Psychological Effects of Wearables 166
3. Social Implications 168
4. Technology Acceptance Factors............................170
References 172

Chapter Seven: Health Concerns 175

1. Introduction...175
2. Electromagnetic Radiation and Specific Absorption Rate 176
3. Thermal Effects..180
4. Health Concerns...180
    4.1 Cancer 180
    4.2 Fertility...181
    4.3 Vision and Sleep Disorders 182
    4.4 Pain and Discomfort 183
    4.5 Other Risks 184
    4.6 Recommendations..185
5. Regulations 186
References...189

Chapter Eight: The Metaverse 191

1. Introduction 191
2. Metaverse Characteristics 194
3. The Elements that Make Up the Metaverse 196
    3.1 Hardware 130
    3.2 Networking Infrastructure 197
    3.3 Platforms 197
    3.4 Protocols and Standards 197
    3.5 Economic and Financial Systems......................................197
    3.5.1 Blockchain...198
    3.5.2 Non Fungible Tokens 200
    3.6 Content and Services 201
4. Applications 201
    4.1 Workplace 202
    4.2 Fashion 204
    4.3 Shopping 204
    4.4 Social Networks/Entertainment 205
    4.5 Tourism 207
    4.6 Healthcare 208
    4.7 Military 208
    4.8 Real Estate 209
    4.9 Manufacturing, Training, and Occupational Safety...210
    4.10 Education 211

4.11 Intimate Relationship 212
5. Concerns and Technical Challenges............................214
    5.1 Concerns 214
    5.1.1 Privacy 214
    5.1.2 Addiction...214
    5.1.3 User Safety...215
    5.1.4 Social Issues..216
    5.1.5 Identity and Reputation Concerns.......................216
    5.1.6 Security 216
    5.1.7 Financial System..217
    5.1.8 Regulation and Legal Issues 217
    5.2 Technical Challenges 217
    5.2.1 Hardware...217
    5.2.2 Networking 219
    5.2.2.1 Bandwidth 219
    5.2.2.2 Latency...220
    5.2.2.3 Reliability..220
Conclusion 221
References...222

CHAPTER ONE

INTRODUCTION AND HISTORICAL BACKGROUND

1. Introduction

Wearables are recognized as one of the hottest trends in today's world. The number of wearable devices available in the market including activity trackers, smart watches, medical aids, and smart fabrics seems to grow exponentially.

Wearable devices have managed to swiftly gain a notable position in the consumer electronics market, and are now making their way to become the new go-to technology to address the needs of many industries. For instance, the construction and mining sectors are increasingly investing in the use of wearable devices for hazard and health management by monitoring the environmental quality, detection of approaching hazards, and assessment of physiological parameters of workers. Moreover, wearables are emerging as a solution to make healthcare accessible in remote areas, and a plethora of wearable devices are already being used by medical professionals to aggregate physiological, behavioral, and biochemical data for diagnosing, treating, and managing chronic diseases.

Wearable technology is all about enabling connectivity amongst users and objects and unobtrusively delivering information and services to the right person at the right time. Their potential benefits are virtually limitless, and their applications are radically changing the way we live, and are opening new opportunities for growth and innovation. This is just the tip of a massive iceberg. In fact, an extremely hot topic of conversation currently is "the Metaverse". What makes this subject serious is that giants like

Facebook and Microsoft are diving in. A Metaverse is a hypothesized iteration of the Internet, supported by 3D virtual environments as well as virtual and augmented reality headsets. Obviously, without Wearable Technology to bridge between the physical and virtual worlds, the metaverse will be nothing but an unattainable fantasy.

In this chapter, a brief history and a general overview of this spectacularly enthralling technology are presented.

Figure 1: A user wearing a Virtual Reality headset to experience a Metaverse environment

1.1 What Exactly is a Wearable Device?

The term "wearable devices" generally refers to electronic and computing technologies that are incorporated into accessories or garments which can comfortably be worn on the user's body. These devices are capable of performing several of the tasks and functions as smartphones, laptops, and tablets. However, in some cases, wearable devices can perform tasks more conveniently, and more efficiently than portable and hand-held devices. They also tend to be more sophisticated in terms of sensory feedback and actuating capabilities as compared to hand-held and portable technologies. The ultimate purpose of wearable technology is to conveniently deliver reliable, consistent, seamless, and hands-free digital services.

Typically, wearable devices provide feedback communications of some sort to allow the user to view/access information in real time. A friendly user interface is also an essential feature of such devices, so is an ergonomic design. Examples of wearable devices include smart watches, bracelets, eyewear (i.e.: glasses, contact lenses), headgears (i.e.: helmets), and smart clothing. Figure 1 depicts the most important possible forms of wearable devices.

Another wearable gadget trending recently is the smart ring. This wearable device is becoming popular among businessmen and individuals who spend a big percentage of their time in meetings and want to get notifications without attracting much attention by glancing at their smartwatch and other mobile devices. Additionally, this gadget can be handy while shopping since the user can use it to swipe for payment or gain access to their vehicle and other smart home appliances. The recently trending rings include; the Amazon Echo Loop which is a smart ring that lets you control Alexa with a single tap, the Lycos Life, Blinq, NFC Opn, and Oura smart rings. With these wearables, the user can keep track of biometric activity data like the number of steps the calories burnt. Furthermore, seniors can activate the panic button which will trigger the ring to send emergency help.

While typical wearable devices tend to refer to items which can be placed external to the body surface or clothing, there are more invasive forms as in the case of implantable electronics and sensors. In the author's opinion, invasive implantables, i.e.: ingestible sensors, under the skin microchips, and smart tattoos, which are generally used for medical and tracking purposes, should not be categorized as wearables since they have different mechanisms and operation requirements. The reader should seek other resources which are dedicated to discuss such devices.

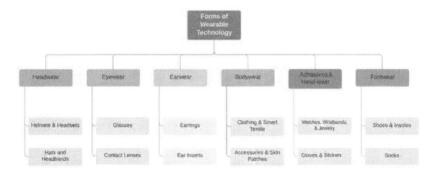

Figure 2: Forms of Wearable Technology

1.2 Characteristics of Wearable Devices

The uses of wearables are numerous and have exciting potentials in the fields of medicine, wellbeing, sports, aging, disabilities, education, transportation, industry, and entertainment. The main objective of wearable technology in each of these fields is to smoothly incorporate functional and portable electronics into the users' daily routines. Prior to their existence in the consumer market, wearables were primarily employed in the field of military technology and health sector.

Generally speaking, wearables share many aspects of the sensing, connectivity, automation, and intelligence characteristics with

Internet of Things devices. However, there are a few major differences worth highlighting which will be discussed in the following sections.

Form factor is a hardware design aspect in electronics packaging which specifies the physical dimensions, shape, weight, and other components specifications of the printed circuit board (PCB) or the device itself. Although wearable devices have a small form factor in general, it is practically dependent on the type and the way they are worn (rings and wristbands, as opposed to glasses and clothing).

Smaller form factors may offer reduced usage of material, easy handling, and simpler logistics; however, they typically give rise to higher design and manufacturing costs in addition to signal integrity issues and maintenance constraints.

Moreover, durability, comfort, aesthetics, and ergonomic factors are important when it comes to designing a wearable device. Weight, shape, color, and texture must be carefully considered. The general characteristics of wearable technology are presented in Figure 2.

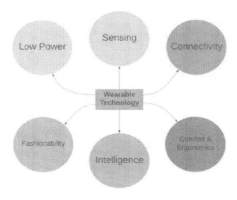

Figure 3: Characteristics of Wearable Technology

1.3 IoT vs. Wearables

We can define the IoT as a network of connected devices with unique identifiers, which have embedded and/or equipped with technologies that enable them to perceive, aggregate, and communicate meaningful information about the environment in which they are placed in, and/or themselves. Despite several commonalities, there are substantial differences when we speak about wearable technology in the context of fitness trackers as opposed to when IoT is used in manufacturing processes or smart cities. In fact, many experts in the field argue that wearables fall under the umbrella of IoT. One key difference worth highlighting here is that a wearable device is not required to have a unique identity (IP address). Moreover, most wearables rely on a gateway device, such as a smartphone, for configuration and connectivity, and in most cases to enable features and process data. This aspect that makes wearables a separate class of devices, and that's why we prefer to treat these as two technologies with two sets of characteristics.

It is also worth noting that not all wearable devices require connectivity, for example, a simple pedometer and an ultraviolet monitor could operate offline. Other wearables require minimal connectivity only.

Although IoT and wearable devices share a lot in common in terms of design aspects, components, and technologies and protocols used, there are still some real differences that architects and designers need to be aware of. Figure 3 shows a table summarizing the main differences between IoT and Wearable Technology.

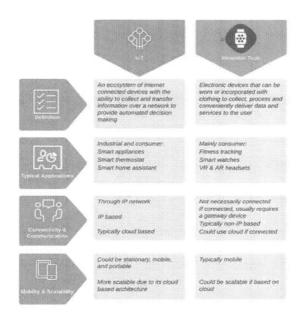

Figure 4: A summary of the main differences between IoT and Wearable Technology

2. A Brief History of Wearable Technology

The beginning of this decade has surely witnessed the increasing number of wearable devices where one can spot numerous variations of smart watches, health assistive gadgets, fitness trackers, and smart clothes on the shelves. The growing number of these sleek devices since then along with their expanding applications clearly indicates that wearables are thriving. But one may ask: When and how did it all begin?

Here, we are not discussing the first ubiquitous wearable technology: the eyeglasses, which dates back to the thirteenth century, nor the abacus ring which dates back to the early days of China's Qing dynasty in the seventeenth century. We are specifically addressing smart wearables that have digital computational power!

You may be surprised to learn that much of the history of this technology is found in a wearable device used to cheat roulette dealers in casinos! In 1961, two mathematicians: Edward Thorp and Claude Shannon built "smart" timing computing devices to help them cheat at the roulette table. This apparatus which was not illegal at the time of invention, could predict where balls would land on a roulette wheel, improving the chance of winning a bet by up to 44%. One of devices was concealed in a shoe (think smart wearable!), while the other in a pack of cigarettes. It is worth noting that Thorp has referred to himself as the inventor of the first wearable computing device. Similar versions of this apparatus were designed and built in the 60s and 70s of the past century targeting the gambling businesses, perhaps the most popular is a shoe-based wearable computer named "George" designed by Keith Taft and operated by his toes which was used to gain an advantage at Blackjack tables.[1]

Prior to this, in 1938, Aurex Corp., a Chicago-based electronics manufacturer developed the first "analog" wearable hearing aid

---

[1] **Claude Shannon** is also known as the father of information theory with his legendary paper "A Mathematical Theory of Communication" published in 1948. He is also well known for founding the digital circuit design and cryptanalysis theories in 1930's when he was in his early 20's as a master's student at the Massachusetts Institute of Technology (MIT).

Edward Oakley "Ed" Thorp is an American mathematics professor at the University of California, Irvine between 1965 and 1977, author of the books (Beat the Dealer and Beat the Market), and blackjack player; he is best known as the "father of the wearable computer".

device, fingerprinting one of the first achievements in the biomedical wearable tech industry. In 1958, the world's first wearable pacemaker was invented by Earl Bakken. One might argue that these are not "smart technologies" since they are not based on a computing system, however, they are still categorized under "technology".

On the other hand, the first "smart watch" was first released in 1975 under the brand "Pulsar". The smart watch was a wearable calculator in essence, which became a widely adopted gadget by science geeks and mathematics nerds across the world. These calculator watches had their heyday through the 80s of the past century. Although their popularity declined drastically, calculator watches are still produced by many companies to this day.

Figure 5: The Pulsar calculator LED watch released in 1975. Photo courtesy of (Peter Crazywatch)

Some might argue that the first mainstream wearable technology ever (in the 1980's through the mid 1990's) was the Walkman music player. The Japanese brand SONY launched this product in 1979 which was followed by an immediate commercial success as it drastically transformed the music listening habits of many people around the world. SONY's Walkman production line was terminated permanently a few years ago with over 220 million copies sold worldwide.

In 1981, wearable computers switched to become more general-purpose when a high school student, Steve Mann, incorporated an Apple II (6502) computer within a steel-framed backpack to control photography equipment attached to a helmet. It is also worth noting that Mann has pioneered many innovations in the fields of wearable computing and digital photography. He is also famous for creating the first wearable wireless webcam in 1994 and as the first lifelogger. [2]

In the healthcare sector, the first practical, wearable, fully digital hearing aid was invented by Engebretson, Morley and Popelka. Their US-registered patent, "Hearing aids, signal supplying apparatus, systems for compensating hearing deficiencies, and methods" filed in 1984 served as the foundation of all subsequent digital hearing aids, including those produced today.

The mid 1990s marked the brainstorming period for wearable technology where conferences and expos on wearables and smart textiles began to see a rise in popularity. The Defense Advanced Research Project Agency (DARPA) held its forward-thinking workshop in 1996 entitled "Wearables in 2005". One of DARPA's

---

[2] *A **lifelogger** typically wears a recording instrument or a computer in order to capture large portions of his/her life.*

galvanizing predictions included computerized gloves that could read RFID tags. However, wearables were overshadowed by the *smartphone revolution* between the late 1990s and mid-2000s, smartphones were simply the consumer's gadget of choice, due to obvious reasons.

Figure 6: Steve Mann wearing one of his wireless wearable webcam
(Photo courtesy of Alchetron)

In more recent years, the wearable technology started to evolve with a mind-blowing speed. In 2002, in demonstration of Kevin Warwick's Project Cyborg, a necklace, worn by his wife, exhibited color changes dependent on signals from Warwick's nervous system. The necklace was electronically connected to Warwick's nervous system via an implanted electrode array.

In 2003, the Garmin Forerunner, a performance-tracking watch, emerged and was followed by the nowadays-popular devices such as the Nike+iPod, Jawbone, and Fitbit activity trackers.

In the late 2000s, a plethora of Chinese companies started producing GSM mobile phones integrated within wristwatches and equipped with mini 1.8-inch displays; while the first smart watch, Pebble, came on the scene in 2012, followed by the much-hyped Apple Watch in 2014.

Future wearable technology devices may serve different functions that one could not yet imagine, but it is clear to see how early wearables evolved into the life-improving devices we enjoy today. In fact, the notion of the Metaverse virtual worlds, for the most part powered by wearable technology, has expanded well beyond the field of immersive gaming, and it is increasingly finding its way into fields such as economics, communications, business, education, and creative industries.

3. The Wearables We Know Today

Perhaps one of the most publicized wearable technologies today is the Apple Watch (a line of smartwatches developed by Apple Inc.). The watch incorporates activity-tracking and health-related capabilities with other Apple applications and products. The main goal of the Apple Watch was to enhance the way users interact with their iPhones. The Apple Watch started when Kevin Lynch was hired by Apple to create a wearable technology for the wrist. He said: "People are carrying their phones with them and looking at the screen so much. People want that level of engagement. But how do we provide it in a way that's a little more human, a little more in the moment when you're with somebody?". The Apple Watch functions when connected to the iPhone via Bluetooth, and by accessing the Watch compatible apps stored on the iPhone.

Figure 7: The Apple Watch (Photo courtesy of Apple Inc.)

Resonating with today's technological advancement, modern consumers take an active role in utilizing wearables to track and record data of their active lifestyles. Nowadays' wearable fitness and health trackers are capable of monitoring the user's biometric data including heart rate, blood pressure, temperature, calories, and sleep patterns.

Another hot wearable: Fitbit, is capable of measuring personal fitness metrics such as the number of steps walked or climbed, heart rate, sleep patterns, and even stress levels.

Figure 8: Fitbit Surge smart watch fitness tracker (Photo courtesy of Fitbit©)

On the other hand, many argues that the single most innovative wearable device of all time is the Google Glass, which in essence is a pair of glasses equipped with a built-in computer and peripherals such as a mini display represented by a 640×360 pixel prism projector that beams the screen into the user's right eye, a touch pad that enables gesture control, a camera, and a microphone. The Glass runs a specialized operating system called Glass OS, and has 2GB of RAM and 16GB of flash storage, in addition to an accelerometer, a gyroscope and a light sensor. Through these essential peripherals, a user can access information by actively interacting with the web, and capture real time activities. The Glass also utilizes voice recognition technology to have the Glass type messages or send requests. It accesses the Internet through two wireless technologies: Wi-Fi and Bluetooth which are linked to the wireless service of the user's mobile phone.

One can imagine a vast number of applications of this technology. In fact, the Glass is already being utilized in a number of very interesting applications that were once considered 'futuristic'. For example, some medical doctors are already utilizing the Glass during surgeries to have a convenient access to the patient's vitals like heart rate, blood pressure, temperature, and Electrocardiogram (ECG) signals.

Obviously, such technology could greatly transform the lives of people with disabilities. For instance, one application is designed to enhance the communication between parents and their deaf children. It allows parents to swiftly look up sign language dictionary via voice commands in order to communicate effectively, instead of having to turn to a book or a computer.

Google released the consumer version of Glass in 2013 amid much fanfare, but it failed to gain commercial success. The Glass also faced serious criticism due to concerns that its use could violate current privacy laws. In 2017, Google launched the Glass Enterprise Edition after deciding that the Glass was better suited to workers who need hands-free access to information, such as in healthcare, manufacturing, and logistics. In 2019, Google has announced a new version of its Enterprise Edition which has an improved processor, camera, charging unit, and various other updates.

One can imagine a considerable number of applications this technology is capable of creating. In fact, the Glass is already being utilized in a number of areas once considered 'futuristic'. For example, Augmedix, a San Francisco based company, developed a Glass app that allows physicians to live-stream the patient visit. The company claims that electronic health record problems will be eliminated, and their system would possibly save doctors up to 15 hours a week.

In 2013, Rafael Grossmann was the first surgeon to demonstrate the use of Google Glass during a live surgical procedure. In the same year, the Glass was used by an Ohio State University surgeon to consult with another colleague, remotely.

Obviously, such technology could have a positive impact on the lives of people with disabilities. For example, one application is designed to enable parents to swiftly access sign language dictionary through voice commands in order to communicate effectively with their deaf children.

Using a smart glass technology in the tourism and leisure industry, the experience of tourists could be substantially improved. Attractions and museum tours can be immensely enhanced by displaying text or providing audible information when recognizable buildings, sculptures, and artwork are detected. Users will also be able to capture photographs and videos more conveniently, i.e.: via voice command or a wink of an eye. Another helpful application dedicated to break the language barriers when travelling provides instantaneous translation. Any text visible to the Glass field of view can be translated via voice commands.

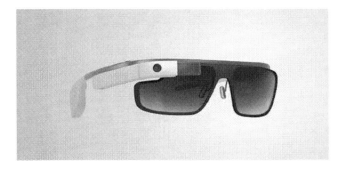

Figure 9: Explorer edition of Google Glass © (Photo courtesy of Google Inc.)

Boeing is using the Glass to help their assembly crew in the connecting aircraft wire harnesses, which is a very lengthy process that requires a high volume of paperwork. The crew now could have a hands-free access to the needed information using voice commands.

Stanford University is conducting a breakthrough research dedicated to help autism patients read the emotions of others using the Glass by utilizing facial recognition software to determine the emotions expressed on the people's faces projected within the display.

In 2014, Novartis and Google X (now X)[3] started to employ wearable technology in the field of telemedicine[4] through the testing of a smart contact lens, namely, the Google Contact Lens (carried out by Verily (formerly Google Life Sciences)). The smart lens is equipped with a miniaturized glucose sensor that continuously measures glucose levels in tears and communicates the data to a smart phone through a wireless microchip to provide glucose information for diabetic users. On November 16, 2018, Verily announced it has discontinued the project because of the lack of correlation between tear glucose and blood glucose. Despite its failure, the project is an eye opener to what wearable technology is capable of.

More recently, adhesives are gaining popularity for use in wearable electronic devices because they provide good structural integrity, impact performance, thermal stability, and resistance to moisture which can play a critical role in a wearable device's overall success in a variety of applications.

In summary, the applications of wearable technology are extremely powerful and virtually unlimited. It is one technology that will revolutionize many aspects of our lives.

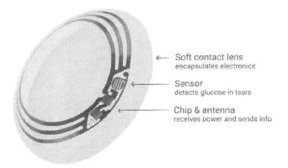

Figure 10: Infographic photo of the Google Smart Lens © (Photo courtesy of Google Inc.)[3]

4. Challenges

While the IoT and Wearable Technology continue to transform our lives in the 21st century, significant challenges that could

---

[3] *X is an American semi-secret research-and-development center founded by Google in 2010 with the name Google X. The company started with the development of Google's self-driving car. It is located about 0.5 mile from Google's headquarters in Mountain View, California. Google X was renamed to X after the restructuring of Google into Alphabet.*

[4] ***Telemedicine***, *in a nutshell, is providing clinical health care from a distance through the use of telecommunication and information technology. It helps to minimize the distance barriers and can greatly improve access to healthcare services that would often not be readily available, especially in rural regions.*

stand in the way of realizing its full potential are coming to light. Below are the major challenges that require full attention:

4.1. Security

Security is one of the cornerstones of the Internet and is the most significant challenge for IoT and wearable devices. The hacking of fitness trackers, security cameras, baby monitors, and other abuses have drawn the attention and serious concerns of major tech firms and governmental agencies across the world.

While security considerations are not new in the realm of information technology, the characteristics of many IoT and wearable technology deployments introduce new and unprecedented security challenges. Addressing these challenges and ensuring secure IoT and wearable products and services must be a top priority. As these technologies are becoming more pervasive and integrated into our daily lives, users need to be assured that these devices and associated data are secure from vulnerabilities such as cyber-attacks and data exposure.

The more consequential shift in security will come from the fact that IoT and wearable technology will become more integrated into our daily activities. Concerns will no longer be limited to the protecting our sensitive data and assets. Our own lives and health can become the target of malicious attacks.

This challenge is further amplified by other considerations such as the mass-scale production of identical devices, the ability of some devices to be automatically paired with other devices, and the potential deployment of these devices in unsecure environments.

4.2. Privacy

While many of the emerging IoT and wearable technologies are giving rise to a spectrum of new applications and innovative uses, as well as promising and attractive benefits, they also pose privacy concerns that are largely unexplored. In fact, a new research area concerning the security and privacy of these technologies has recently emerged. Additionally, the need for the majority of wearable devices to interact and share data with an access point (i.e.: a smart watch to smart phone, medical monitoring device to a home server, etc.) along with other sensors and peripherals would certainly create a new class of privacy and security hazards.

Some IoT and wearable devices deploy various sensors to collect a wide spectrum of biological, environmental, behavioral and social information from and for their users. Clearly, the more these devices are incorporated into our daily lives, the greater the amount of sensitive information will be transported, stored, and processed by these devices, which also elevate privacy concerns.

Moreover, integrated voice recognition or monitoring features are continuously listening to conversations; video record activities and selectively transmit such potentially sensitive data to a cloud service for processing, which sometimes involves a third party. Handling and interacting with such information unveil legal and regulatory challenges facing data protection and privacy laws.

One specific privacy concern associated with the emerging smart glasses is that they allow users to simultaneously record and share images and videos of people and their activities in their range of vision, in real time. This problem will soon be intensified when such devices are integrated with facial recognition programs which will allow users to see the person's name in the field of view, personal information, and even visit their social media accounts.

4.3. Standards and Regulations

The lack of standards and best practices documentations poses a major limitation to the potential of IoT and wearable devices. Without standards to guide manufacturers and developers, these products that often operate in a disruptive manner, would lead to interoperability issues, and might have negative impacts if poorly designed and configured. Such devices can have adverse consequences on the network resources they connect to and the broader Internet. Unfortunately, most of this comes down to cost constraints and the pressuring need to release a product to the market quicker than competitors. Moreover, there is a wide range of regulatory and legal questions surrounding the IoT and wearable technology, which require thoughtful consideration.

Legal issues with IoT and wearable devices may include conflicts between governmental surveillance and civil rights; policies of data retention and destruction; legal liability/penalty for unintended uses, and security breaches or privacy abuses. Furthermore, technology is advancing much faster than the associated policy and regulatory environments which might render policies and regulations to be inappropriate.

Big data presents another serious challenge. The analysis, extraction, manipulation, storage and processing of substantial amounts of data may pose other legal problems as in profiling, behavior analysis and monitoring. Big data may require new protection policies, international coordination, and infrastructure management, among others.

Furthermore, the cloud and even the Internet itself are not tied to one specific geographic location. Moreover, the sheer amount of IoT and wearable devices originate from a number of different sources, including international partners and vendors, which makes it impossible for a localized regulatory authority to enforce quality control or standardized tests.

As of today, these challenges have been minimally acted upon by policy makers. However, they reflect a pressing necessity to seek government solutions to both pronounce the strengths of these technologies and deploy policies to minimize their risks.

4.4. Energy and Power Issues

The increase in data rates, and the number of Internet-enabled services and the exponential growth of IoT and wearable devices are leading to a substantial increase in network energy consumption.

Moreover, the push toward smaller size and lower power is creating more signal and power integrity problems in wearable devices. Common issues include mutual coupling, distortion, excessive losses, impedance mismatch, and generator noise. Failure to deal with these issues can have detrimental effects on these devices.

4.5. Connectivity

According to recent research reports, around 31 billion IoT and wearable devices will be connected to the Internet by 2025. Thus, it's just a matter of time before users begin to experience substantial bottlenecks in connectivity, proficiency, and overall performance.

Currently, a big percentage of connected devices rely on centralized, and server/client platforms to authenticate, authorize, and connect additional nodes in a given network. This model is sufficient for now but as additional billions of devices join the network, such platforms will turn into a bottleneck. Such systems will require improved cloud servers that can handle such large amounts of information traffic. This is already being addressed by the academic and industrial community which is pushing towards decentralized networks. With such networks, some of the tasks are moved to the edge, such as using fog computing, which takes charge of time-sensitive operations (this will be discussed in detail in the following chapters), whereas cloud servers take on data assembly and analytical responsibilities.

Conclusion

Wearable devices are enabled by the latest developments in smart sensors, embedded systems, and communication technologies and protocols. The fundamental premise is to have sensors and actuators work autonomously to deliver a new class of applications. The recent technological revolution gave rise to the first phase of wearable devices, and in the next few years, these devices are expected to bridge diverse technologies to enable novel applications.

Benefits are substantial, but so are the challenges. This will require businesses, governments, standards bodies, and academia to work together toward a common goal.

References

[1] High Tech Casino Advantage Play: Legislative Approaches to the Threat of Predictive Devices, David W. Schnell-Davis, University of Nevada, Las Vegas Gaming Law Journal, Vol. 3, P. 299-346, Fall, 2012.

[2] E.O. Thorp, "Optimal Gambling Systems for Favorable Games," Review of the International Statistical Institute, Vol. 37, 1969, pp. 273-293.

[3] E.O. Thorp, "Systems for Roulette I," Gambling Times, January/February 1979.

[4] E.O. Thorp, The Mathematics of Gambling, Lyle Stuart, Secaucus, New Jersey, 1984.

[5] National Institute for Empowerment of Persons with Multiple Disabilities (Manual), ISBN: 978-81-928032-1-0, 2014.

[6] Wearable Technology and Mobile Innovations for Next-Generation Education, Janet Holland, IGI Global, ISBN-13:9781522500698, 2016.

[7] Bertolucci, J., Reliability report card: grading tech's biggest brands. PC World, 27(2). Chan, J., 2010, November 4.

[8] A brief history of wearable computing, Bradley Rhodes - MIT Media Lab, MIT Wearable Computing Project, https://www.media.mit.edu/wearables/lizzy/timeline.html, last accessed: January, 2017.

[9] A, Erfinder A. Maynard Engebretson, Robert E. Morley, Jr., Gerald R. Popelka, Hearing aids, signal supplying apparatus, systems for compensating hearing deficiencies, and methods, US patent 4548082.

[10] Wearable Computer Applications A Future Perspective, Kalpesh A. Popat, Dr. Priyanka Sharma, International Journal of Engineering and Innovative Technology (IJEIT), Volume 3, Issue 1, July 2013.

[11] Innovation in Wearable and Flexible Antennas (book), Haider Raad Khaleel, WIT Press, 2014.

[12] The Wearable Technology Handbook, Haider Raad, United Scholars Publications, 2017.

[13] The Sensor Cloud the Homeland Security, http://www.mistralsolutions.com/hs-downloads/tech-briefs/nov11-article3.html, 2011.

[14] O. Vermesan, P. Friess, P. Guillemin et al., "Internet of things strategic research roadmap," in Internet of Things: Global Technological and Societal Trends, vol. 1, pp. 9–52, 2011.

[15] I. Pe~na-Lopez, Itu Internet Report 2005: The Internet of Things, 2005.

[16] I. Mashal, O. Alsaryrah, T.-Y. Chung, C.-Z. Yang, W.-H. Kuo, and D. P. Agrawal, "Choices for interaction with things on Internet and underlying issues," Ad Hoc Networks, vol. 28, pp. 68–90, 2015.

[17] O. Said and M. Masud, "Towards internet of things: survey and future vision," International Journal of Computer Networks, vol. 5, no. 1, pp. 1–17, 2013.

[18] M. Aazam and E.-N. Huh, "Fog computing and smart gateway-based communication for cloud of things," in Proceedings of the 2nd IEEE International Conference on Future Internet of Things and Cloud (FiCloud '14), pp. 464–470, Barcelona, Spain, August 2014.

[19] L. Atzori, A. Iera, and G. Morabito, "SIoT: giving a social structure to the internet of things," IEEE Communications Letters, vol. 15, no. 11, pp. 1193–1195, 2011.

[20] Z. Sheng, S. Yang, Y. Yu, A. Vasilakos, J. Mccann, and K. Leung, "A survey on the ietf protocol suite for the internet of things: standards, challenges, and opportunities," IEEE Wireless Communications, vol. 20, no. 6, pp. 91–98, 2013.

[21] B. Guo, D. Zhang, Z. Wang, Z. Yu, and X. Zhou, "Opportunistic IoT: Exploring the harmonious interaction between human and the internet of things," Journal of Network and Computer Applications, vol. 36, no. 6, pp. 1531–1539, 2013.

[22] G. Liang, J. Cao, and W. Zhu, "CircleSense: a pervasive computing system for recognizing social activities," in Proceedings of the 11th IEEE International Conference on Pervasive Computing and Communications (PerCom '13), pp. 201–206, IEEE, San Diego, Calif, USA, March 2013.

[23] High-Performance Big Data Analytics: The Solution Approaches and Systems, Springer-Verlag, London, http://www.springer.com/in/book/9783319207438, November 2015.

CHAPTER TWO

APPLICATIONS OF WEARABLE TECHNOLOGY

1. Introduction

As an emerging industry, wearables have made the integration of technology into many facets of our daily lives possible. Not too long ago, such technology integrations were considered merely science fiction! This chapter covers the applications of wearables in various fields. It also provides an insight on the roles this technology could play in practice and discusses the challenges and key success factors for its adoption. It should be noted that applications of the Metaverse will be discussed in detail in Chapter 9.

2. Applications

2.1 Biomedicine and Healthcare

As mentioned previously, Aurex Corp. developed the first wearable hearing aid device in 1938, which is considered as one of the early milestones in the biomedical wearable tech industry.

Nowadays, healthcare wearables firms tend to target the most common chronic diseases: diabetes, congestive heart failure, hypertension, and COPD (chronic obstructive pulmonary disease), in addition to pain management. For example, Vital Connect Inc., a California based medical tech company developed the HealthPatch MD, which is a biosensor that monitors the patient's heart and respiratory rates, temperature, in addition to body posture. It is also equipped with a fall detection capability and can be used at home or in a hospital setting.

Abbott Healthcare developed a wearable continuous glucose monitoring device for diabetic patients, with the brand name FreeStyle Libre Flash. The FDA approved the MiniMed 670G System developed by Medtronic, a hybrid closed loop system that monitors and adjusts glucose levels by automatically dispensing basal insulin replicating some of the functions of an actual pancreas.

iRhythm technologies Inc, on the other hand, developed the Zio XT Patch, which functions toward the detection of abnormal heart activity. The device is based on water-proof electrocardiogram patches which are continuously worn for two weeks. The collected data is then forwarded to iRhythm's clinical app for processing.

In the pain management area, NeuroMetrix developed a wearable device called Quell, which is an FDA approved device geared toward reducing pain. This device utilizes an accelerometer to compute and assess a user's activity level and trigger its stimulation intensity accordingly to attenuate pain. Moreover, the device is equipped with Bluetooth technology to sync to a smartphone app, where a patient can control the device's settings and track visual results of therapy and sleep.

The WristOx2 developed by Nonin is a pulse oximeter which is geared toward patients with asthma and COPD. The device can be used in a hospital setting or at home after discharge to allow remote and extended monitoring of the patient's heart rate and oxygen levels.

Figure 1: MiniMed 670G glucose level monitoring and adjustment system. Photo courtesy of Medtronic

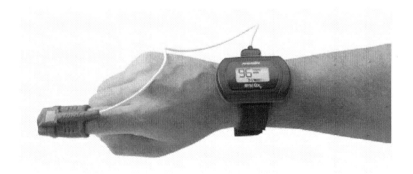

Figure 2: Nonin Medical's WristOx$_2$® Model 3150 wrist-worn pulse oximeter. Photo courtesy of Nonin

Monica AN24 is compact wearable device that enables accurate fetal monitoring for patients with high Body Mass Index (BMI) patients. The device is wireless and does not require any belts which improves the patients' comfort during the baby delivery process. While the OvulaRing, developed by VivoSensMedical GmbH, is designed to assist women who are trying to conceive in determining when they are ovulating and most fertile. The wearable device is inserted into the vagina where the biosensor monitors the basal body temperature throughout the menstrual cycle. [4]

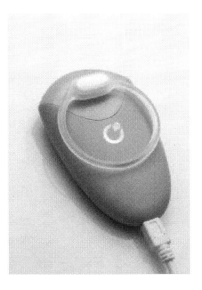

Figure 3: OvulaRing, Photo courtesy of VivoSensMedical GmbH

---

[4] *An **oximeter** is a noninvasive device for monitoring a user's oxygen saturation levels.*

Telemedicine is broadly defined as providing medical and healthcare services through telecommunications to enhance the patient's health conditions. Information, computing, and telecommunication technologies have enabled a wide range of telemedicine applications. The first use of telemedicine dates back to the invention of the telephone in 1876, with medical consultations provided over the phone by physicians. Evidently, the recent groundbreaking advances in wearable technology have brought medical care to virtually all corners of the world.

With telemedicine, a user can access medical care that may not be available locally. Thus, the transportation, geographic, and cultural barriers to access high quality healthcare are minimized. For example, a surgeon can nowadays perform surgeries and track the progress of the post-surgical recovery remotely.

Theoretically, wearable technology can be employed in a surgical practice in three main aspects: assistance, assessment, and augmentation.

Assistance refers to the use of wearable technologies to compensate a physical task during a surgery. For example, the use of an arm-mounted device allowing gesture control of a picture archiving and communication system (PACS) to provide cross-sectional imaging during surgeries without violating the sterility standards. Assessment, on the other hand, points out to the utilization of wearables in the quantitative measurement of illness severity, clinical outcomes, or in surgical education, while augmentation deals with the real-time access of information by the surgeon in clinical and surgical settings.

It is also widely acknowledged that cultural and social barriers may prevent patients from seeking necessary mental health services in many parts of the world. Recent studies have confirmed that telemedicine is very effective in overcoming such barriers, the subcategory is known as tele-psychiatry.

On the other hand, the intersection between telemedicine and pet wearables is represented by the PetPace collar which is an activity and wellness tracker designed for dogs and cats. The device links to a network service where it can be accessed by veterinary systems. PetPace collar is capable of tracking a pet's heart and respiration rates, temperature, activity levels, and much more.[5]

Figure 4: PetPace collar tracker worn by a cat. Photo courtesy of PetPace

---

[5] ***PACS*** *is a medical imaging technology which offers economical storage and instant access to images from various models.*

2.1.1. Adhesives

The wearable medical devices market will further evolve as patients and consumers continue to keep track of their health parameters. Researchers are trying to find new ways to improve non-pharmaceutical therapies and digital monitoring. The objective of these efforts is to increase the quality of life and longevity of the aging population living at home and enhance virtual care. The industry holy grail is to offer progressively effective, personalized monitoring of health and care delivery.

To address this growing need, tech manufacturers are working to find more innovative methods to miniaturize these devices and making them lighter and less invasive. To achieve some of these advancements, one solution is adhesives. Adhesives can play a vital role in a wearable electronic device's overall success, whether it's bond device components or adhering devices to skin. Hence, it's crucial that adhesives evolve along with wearable medical devices to cultivate continued innovation.

Hiking, power exercise, athletic training, showering and swimming are all activities users should be able to enjoy without their personal device hindering them. Situations in which a device may come into contact with sweat or other moistures and fluids pose a great deal of challenge to design engineers; resilience, durability against wear and tear is crucial. Moreover, comfort and aesthetics are of paramount importance for users required to use a wearable device long-term, such as continuous glucose monitors or heart rate monitors. Having adhesive solutions that are able to handle everyday life use, will help allow devices to stay connected with the user longer.

Although adhesives are a tiny component of wearables, they're often one of the champions of medical device innovation. As the medical wearables industry continues to change the way we interact with our health, tech designers and adhesive experts will need to keep working towards producing smarter devices using skin friendly materials.

2.2 Fitness and Wellbeing

The demand on fitness trackers that come packed with different types of technologies is on the rise. Fitbit, Jawbone, Garmin, and Misfit are probably the most popular brand names in today's wearables market. Integrated with sensory and wireless capabilities, such activity trackers measure fitness-oriented personal metrics such as the number of steps climbed or walked, heart rate, sleep and stress patterns, and some have the capability to determine the user's location through a built-in Global Positioning System.

Figure 5: Fitbit Surge used by an athlete. Photo courtesy of Fitbit

A new trend started by many companies is the use of wearable devices to track their employees' health and activity levels by asking them to use fitness monitors as part of a wellbeing program. The acquired data is then forwarded to their health insurance providers which offer reduced policy premiums or some other incentives in return. According to Gartner, an American IT and technology research and advisory firm, around 10,000 companies across the world offered the use of fitness trackers to their employees in 2014.

It is worth mentioning that many researchers agree that placing fitness trackers on the user's hip or foot rather than the wrist would offer more accurate readings. Even successful fitness trackers like Fitbit exaggerate step count in some scenarios and can misinterpret enthusiastic gestures as steps. This fact has triggered some wearable tech companies to develop smart socks, bras, and boxers. For example, Sensoria's Fitness Socks track the user's personal health metrics. The device comprises a cuff-shaped fitness tracker which magnetically connects to Sensoria's running-friendly fabric smart socks. Intuitively, the device communicates with a smart phone app, keeps tabs of the user's activities and guides them via audio cues during running. Sensoria claims that their technology offers higher accuracy in measuring steps, velocity, altitude, and burned calories through their unique foot landing technique and weight distribution on the feet. However, such platforms could potentially pose practicality and convenience issues.

Figure 6: Sensoria's Fitness Socks, photo courtesy of Sensoria

Figure 7: SunFriend sun exposure wristband, photo courtesy of SunFriend

Another offering in the area of wellbeing is the SunFriend, which is a wristwatch that uses ultra violet sensors to track the amount of sunlight exposure received by the body. The sensors data is then visualized via LED lights that start blinking when the wearer's ultra violet levels are within the dangerous zone. Such device would be very useful in countries with high skin cancer rates due to excessive exposure to ultraviolet as in Australia, New Zealand, Argentina, Denmark, and parts of USA.

In response to the COVID-19 pandemic, researchers and scientists all over the world tried to apply their knowledge and expertise to defeat the spread. For example, researchers at Rockefeller Neuroscience Institute reported that data from a smart ring, which is a wearable sleep and activity tracker, can be synced to an app that measures vital signs to predict, in advance, whether a person may develop COVID-19 symptoms. The device was successfully able to predict that a person would experience symptoms such as a cough, fever, and shortness of breath up to three days before they are manifested. On the other hand, Honk Kong started its quarantine efforts at the airport where arriving passengers were given smart wrist-bands to track their movements in the city and enforce their stay in quarantine. This technology is called "Geofencing", where a virtual space is created using GPS, RFID, Wi-Fi, Bluetooth, and cellular network. Another wearable device developed in Italy is based on smart bracelet; when two smart bands are in proximity, they vibrate, thus emitting an alert signal that reminds people to respect social distancing.

2.3 Sports

Wearable technology is changing every aspect of our lives including how we experience sports. Athletes are already using wearables to enhance their performance. On the other side of the arena, devices like Google Glass are giving fans an entirely new perspective of the game.

For example, Adidas is working with professional soccer teams in the west coast to monitor heart rate and other metrics of players using its miCoach wearable technology. The collected data is analyzed by coaches to track the players' performance and have the best estimate on scheduling breaks to minimize the risk of injury. Other wearables used in similar applications are worn underneath the players' clothing and monitor tens of parameters including velocity, orientation, acceleration, blood pressure, and heart rate. The data is then forwarded "wirelessly" to the coach's smartphone.

For everyday athletes, devices like PUSH, OMSignal, and Sensoria provide sports-related biometric information so that they could train like professionals. OMSignal and Sensoria produce bras, training shirts, and socks, all smart! It is worth mentioning that the data harvested from the OMSignal's smart bra or workout shirt is used for more than just workout analysis. Their products can, for example, track stress levels through the integrated heart rate, muscle tension, and breathing pattern sensors, and alert the users via text or push notification to have them aware of their emotional state, thus contributing to the improvement of both mental and physical well-being.

Recon Snow2 is another impressive sports wearable gadget. In essence, it is a powerful heads-up display designed especially for alpine sports with the onboard processing power, sensory, and networking capabilities comparable to that of a smartphone. The Recon Snow2 is intended for skiers and snowboarders to stay connected. Its onboard sensors can provide speed and altitude; social networking, profiles and tracking of other skiers on the resort, and map apps, all hands-free. It simply gives you the power of a smartphone into your field of vision.

Figure 8: OMSignal's smart bra (OMbra), photo courtesy of OMSignal

Figure 9: The Oakley Airwave skier's goggle, photo courtesy of Recon

2.4 Entertainment

Technologies like virtual reality and haptics are redefining the way we experience and make movies and music, and even the way we enjoy sport games.

For example, Oculus Rift, a virtual reality device is being used to create immersive movies that are incomparable to what we have experienced before. While Condition One, a video technology startup, incorporates advanced 3D graphics, super high-resolution video and cognitive storytelling to create new forms of movies optimized for the Oculus Rift.

On the other hand, wearable devices with haptic feedback and gesture control are defining a new way to make and listen to music. For instance, Imogen Heap, an English singer-songwriter and composer, recently unveiled the MiMu Gloves which utilize gesture and motion to create software-controlled digital music. The MiMu gloves are intended to make the experience of making digital music more physical.

In the area of sport games, CrowdOptic is collaborating with professional basketball teams to enhance the fans experience using Google Glass. Their software captures new perspectives of the game in real-time which are immediately shared with fans.

Theme parks around the world are visited by millions every year. With the emergence of wearable technology, some parks started to employ it to enhance the entertainment experience of tourists. For example, MyMagic+ is a wristband provided by Disney to help the tourist navigate their theme parks. The wristband links the tourist information to a database and serves as an admission ticket, hotel key and credit card. The tourist can schedule their visits to each theme park and pre-order their food without waiting in the extremely long lines.

Accesso Prism is another wearable device that allows the tourist to have a wait-free experience by scheduling or changing a ride selection, reserving a place in line, and monitoring your ride return time without the need to visit kiosks.[6]

Figure 10: Oculus Rift worn by a gamer. Photo courtesy of Oculus. The Oculus Rift is a virtual reality head-mounted display geared towards next generation games and movies, created by Oculus VR Inc. in 2013. It is intended to provide the user with an enhanced experience and advanced level of immersion

---

[6] *Haptics*, *also known as kinesthetic communication, reconstructs the sense of touch by applying forces, vibrations, or motions to the user through actuators, mainly for the purpose of interacting with computer applications. By using sensor-based input/output devices, such as a glove or a joystick, users can receive feedback from computer applications in the form of felt sensations.*

2.5 Gaming

According to industry experts, the use of wearables such as wristbands, earbuds, and eyewear will increase exponentially in gaming platforms as control devices for the virtual reality and biometric gaming market. It is also reported that devices such as rings and smartwatches, would be able to accurately track the user's movements and interface with a virtual reality platform to provide an immersive gaming experience. While a user's biometric response would directly affect the dynamics of the game in biometric gaming.

The real success story of wearable technology in virtual reality gaming has been the Oculus Rift. Unlike the unsuccessful previous virtual reality platforms, Oculus Rift has taken gaming to new levels, introducing immersive games that have made the most of the technology.

According to Valencell, an American biometric technology company that develops biometric sensot technology for wearables, a potential application of biometric gaming may include action games that require gamers to hold their breath while the game character is underwater. Another application would be for the gamer's heart rate to directly affect their accuracy in a shooting-based game. As reported by another recent market analysis, Global augmented/virtual reality (AR/VR) market revenue is expected to reach $661.4 billion by 2025; this should give an idea of where this technology is headed.

In the world of online casinos, Microgaming developers swiftly adapted to the mobile technology, allowing online gamblers to play their favorite games whenever and wherever they are. The company is doing the same with the emergence of wearables via developing the first five-reel smartwatch slot game app. The company is also suggesting that it is only a matter of time before the products of major players in the online casino business are accessible through popular wearables such as the Apple Watch.

2.6 Pets

According to a recent study by Grand View Research, Inc., the global revenue of pet wearable market is expected to reach 2.36 billion US dollar by 2022. The growth in pet ownership and expenditure on pets is anticipated to push product demand over the forecast period. Moreover, the raised awareness towards pet health and fitness is expected to drive substantial investment for the research and development of even more sophisticated wearable products for pets.

Smartphone apps along with connected products are allowing owners to remotely track, monitor, and feed their pets. Some advanced technologies have the capability to remotely unlock and lock home doors to allow access for pet sitters and dog walkers.

As of 2016, Whistle, a California based company, is considered a leading maker of wearable devices for pets. Whistle's on-collar tracker tracks data of the physical activity through a mobile app to keep records of their pet's behavior. Such behavioral information allows the owner to determine whether an indicator of potential health problems exists. Recently, Whistle has brought Global Positioning System (GPS) to their newest generation device which allows owners to track the location of their pet in addition to the activity tracking capabilities.

Figure 11: Whistle's Collar product (left), User Interface (Right). Photo courtesy of Whistle

2.7 Military and Public Safety

In military applications, wearable technology is a fundamental component of the connected soldier system that offers a tactical edge, tracking, and safety improvement. Continuous efforts are made by the research and development arena in this field to advance lightweight electronics, miniaturize antennas, and produce more effective radio communication systems with an eye towards improving soldier mobility and enhancing system deployment and scalability.

U.S. Department of Defense continues to investigate and invest in wearable systems for fighters on the network-oriented battlefield. While engineers at the U.S. Army Research and Development center are collaborating with industry to advance wearable electro-optics based devices geared for body area networks. The concept involves a variety of secure physiological sensors that monitor heart rate, breathing patterns, temperature, and blast effects. The data would be available for real-time wireless transmission to head-quarters or for immediate access.

On the other hand, many law-enforcement departments are already using wearable headsets which include a camera, an ear phone, and a microphone. For example, the London Metropolitan Police's deployment of wearable technology aims at rebuilding the public trust in an entity that has been involved in controversial cases following the killing incidents that triggered the London riots in 2011. Despite the raised privacy concerns, the plan seems to have shown success in collecting offences evidence thereby speeding up the justice process.

The author predicts that virtually every police officer will be wearing a camera by 2030, and they will most likely be equipped with a wrist or eye-wear for accessing information, hands-free.

2.8 Civil Defense

Recent studies confirmed that wearable technology can be valuable in field communications and would improve situational awareness in civil defense and public safety applications. This would ultimately promote multitasking and allow a more educated decision making.

For example, Vienna University of Technology developed a wearable device designed to aggregate data captured by firefighters called ProFiTex. The device maps the firefighters' surroundings on the scene by a color-coded representation. This is achieved by wearable cameras and sensors integrated within the firefighter's helmet. The temperature mapping is also capable of identifying the location of people through smoke and help to decide if specific rooms are safe to enter.

2.9 Travel and Tourism

Previous studies have confirmed that portable and mobile technologies such as notebook, tablets, and smartphones play a significant role in the area of travel and tourism. With the emergence of newer technologies, the travel and tourism industries are faced with new challenges stemming from consumers' behavioral changes influenced by these new technologies.

Several deployments of Google Glass in the aviation sector have been recently observed within its explorer program. For instance, the staff of Virgin Atlantic Airlines at Heathrow Airport has used the Google Glass to provide an enhanced customer service in 2014. The Glass enabled Virgin Atlantic's staff to offer timely weather information and language translations to their clients.

The hospitality industry is another sector adopting and making use of the Google Glass. For example, Starwood Hotels developed an app for Google Glass that enables their preferred guests to use virtual room keys instead of physical ones, get directions to the hotel, and access their reservations and star point balances.

2.10 Aerospace

Up to this day, astronauts depend on printed instruction manuals in case of a system error or an emergency. This is proven to be inefficient which ultimately forces the crew to place a call to the ground station for a solution. However, telecommunication becomes impractical the farther a space vehicle travels away from earth. For example, it would take 20 minutes for a message to travel from Mars to a ground station on Earth.

To overcome such problems, the U.S. National Aeronautics and Space Administration is developing smart glasses for astronauts that can guide them through a repair process or conducting an experiment in outer space. While NASA is collaborating with Osterhout Design Group (San Francisco based tech company), which produces augmented-reality glasses that beam information onto the lenses. The project's plan is to build a system where instruction manuals can be uploaded to the smart glasses, enabling astronauts to follow directions, hands free.

Further, the Australian aerospace company (TAE) will make CSIRO's Guardian Mentor Remote (GMR) wearable technology system available to the global aerospace industry. GMR is a hands-free technology that utilizes a headset and glasses to connect experts with local operators and technicians so they can provide real-time assistance which has a great potential to reduce the maintenance costs for airlines.

2.11 Business and Industry

According to the market intelligence firm Tractica, enterprises are beginning to drill into harnessing the benefits of wearable technology in terms of applicability and practicality. Organizations that are taking initiatives in this area range from large hospitals to small clinics and from giant firms to small and medium-sized businesses. For example, an implantable Near Field Communication microchip allows about 20% of employees of Epicenter (a Swedish company) to swipe into their offices, set an alarm system, register loyalty retailers' points and access their fitness center.

It should be noted that wearable devices like smart glasses, wristbands, and badges are more welcomed by employees than implantable ones. The promise of data captured from monitoring staff movements to improve productivity and efficiency might be very tempting to the management, but it comes with a pressing issue: privacy concerns, which will be discussed in Chapter 6 of this book.

In industry settings, the potential market for wearable technology solutions is expected to exceed that of the general consumer market. Firms in the field service industry have already witnessed the impact of wearable technology, with technicians and engineers wearing camera-based headsets while out in the field.

Kopin, a technology provider of the popular wearable headset Golden-i, claims that its device Gen 3.8 headset, geared toward the light industrial sector, will improve worker's productivity, efficiency and safety. While Vuzix, a New York-based company, offers a variety of glasses and headset solutions, and works on innovative solutions for warehouse management systems.

It is worth noting that head-worn devices might not be appropriate for all applications in an industrial setting. Hence, Fujitsu is working on a wearable device designed for industrial maintenance and on-site operations in the form of a glove. The device is integrated with a Near Field Communication tag reader and features a gesture-based input controller.

2.12 Education

Many educators are realizing that the emerging technologies offer an opportunity to enhance learning instead of being a distraction. In fact, researchers have recently confirmed the potential pedagogical uses of wearable technologies.

For more than two decades, Microsoft's PowerPoint has served educators in almost every discipline as a valuable illustrative visual learning tool. However, the next generation of tech-oriented students will most likely enjoy a more immersive class going beyond a simple slideshow.

In response to this, hundreds of classrooms have already deployed wearable platforms to transform the learning experience of students. For example, Google Expeditions, an educational initiative introduced by Google, uses a folded piece of cardboard with a special pair of lenses to turn a smartphone into a virtual reality gadget. This "smart" cardboard can take students to places such as the Great Wall of China, the Grand Canyon in USA, or Verona in Italy, which serves as a great tool for active and hands-on learning. "The creativity we have seen from teachers, and the engagement from students, has been incredible," as reported by Google's product manager for Expeditions.

Wearable technology in education, if used properly, can increase a student's ability to be more creative and innovative. Students can access information more efficiently without distractions. Other examples of wearable technology in the classroom: are: Autographer, which ensures complete note taking in a classroom setting. While Keyglove are smart gloves that are useful in arts, design, music and data entry. GoPro, another noteworthy example, is a portable camera that can capture a student or teacher's perspective on a lesson, a story, or an event.

2.13   Fashion

Some technology experts say that it's only a matter of time before having smart clothes and accessories becomes mainstream and consumers begin to wonder why they would purchase a piece of clothing if it is not smart.

Today, when we discuss the wearable technology topic, the first thing that comes to mind is the plain looking smart watches and activity trackers which, from the fashion experts' perspective, still lack "style". As wearable technology becomes more mature, developers realize that collaborating with fashion designers is a must to create stylish products people want to buy and show off.

One of the first wearable tech-fashion collaboration was between Martian and Guess producing a Guess Connect Smart Watch. While Tory Burch designed various accessory designs that go with Fitbit, one of the "not" best-looking wearables in the market.

On the other hand, the impressive thing about the collaboration between Swarovski and Misfit trackers is that the design actually offers more technology enhancements. The Misfit Shine activity tracker governed by a large single Swarovski crystal, has several accessories to accompany it. More interestingly, there is a solar powered version, which utilizes light refracted by the crystal to power the device, perpetually.

Tag Heuer's first smartwatch released in 2015 was perhaps the biggest collaboration between a watchmaker and Hi-tech companies. The Tag Heuer Connected was a collaborative work of Google and Intel, running Android Wear along with custom Heuer watch skins.

The fashion guru Ralph Lauren has also joined the game with the Polo Tech Shirt from which is claimed to be the ultimate activity tracker. The smart shirt has woven sensors into the fabric that are capable of tracking sport metrics such as heart rate, pressure, temperature and breathing patterns. The reader may guess that this tech shirt comes with a hefty price tag.

The Unseen, a British company, is producing a line of luxury smart garments and accessories for Selfridges, the high end department stores in the United Kingdom. The products include backpacks, scarves, and phone cases which respond to user interaction or the environment factors like air pressure, body temperature, touch, wind and sunlight by changing colors.

2.14　Novel and Unusual Applications

The reader might not think that wearables find a natural use in baby diapers, but Huggies has developed a moisture sensor based device that can be attached to the outside of a baby's diaper called a Tweet Pee. The device links to a parent's smartphone and alerts them when the diaper needs to be changed. While the device is mainly aimed at providing convenience to the parents, there's also a marketing catch to it. Parents can register each box of diapers they purchase on the smart phone app, the information is then used to send parents a notification when their supply is running low.

Sensoree, a smart fabric startup is utilizing Galvanic Skin Response, using sensors placed on the owner's hands to determine their mood. Their smart shirt (called the Mood Sweater) has a collar of LED lights which changes color based on the wearer's mood. While Biolinq (formerly ElectroZyme) have developed a printed electrochemical biosensor which gives the wearer information about blood and biomarkers such as pH and salt levels by applying a temporary tattoo directly on to the user's skin.

Pollution is a major issue in many urban cities, which poses significant health risks. The Diffus Climate Dress might be the first wearable technology aimed at alerting users to avoid places with high levels of pollution. The garment comprises sensors which measure the Carbon Dioxide concentration in the air. Depending on the CO_2 amount, woven LEDS are activated and displays moving patterns ranging from slow to hectic and chaotic. Also, the dress could potentially be exploited to raise awareness.

The Hug Shirt designed by CuteCircuit is a shirt that allows people send hugs to loved ones over distance! The shirt is integrated with sensors that garner the strength, duration, and location of the touch. Moreover, the skin warmth and the heart rate of the sender are recreated by built-in actuators.

Statistics show that one in eight women will develop breast cancer during their lifetime, and in many cases, early detection can be a life saver. Hence, regular self-exam is of paramount importance, in addition to the regular physician visits. However, the majority of women either forget about the self-exam or are unaware of the consequences. Harnessing the power of wearable technology, Nestle` thinks it has an effective solution called the Tweeting Bra. It is activated by the hidden mechanism under the hook of the bra. Every time the bra is unhooked, it sends a signal to the user's smartphone, the phone then notifies a special server which in turn generates the tweet. The tweet, for example, will read: "Jessica has just unhooked her bra. When you do the same, don't forget about your self-exam.".

The Aurora Dreamband developed by iWinks is the first smart wearable device designed to enhance Rapid Eye Movement (REM) dreaming. Aurora tracks the user's brainwaves and body movements while asleep using an accelerometer, gyroscope, EEG, ECG, EMG, and EOG sensors. Awareness and perception are enhanced through a set of actuators once a dream is detected. It can also serve as a smart alarm clock and sync to a smartphone app to track sleep and dreaming related metrics.

Lastly, A crowd-funded Singapore based startup introduced the world's first wearable smart sex toy, Vibease. The device is a vibrator that can be controlled by a smart phone app and Bluetooth connectivity. According to the company spokesman, the product is aimed at providing users with an immersive experience through syncing it with audio books, in addition to enabling an intimate interaction for couples involved in a long-distance relationship (i.e.: allows a partner to control the vibrations remotely!).

Figure 12: The Aurora Dreamband. Photo courtesy of Daniel Schoonover, iWinks

Conclusion

Wearable Technology are ripe for new and creative ideas to add to the applications already in use. It provides a nearly endless supply of opportunities to interconnect our devices and equipment. When it comes to innovation, this field is wide open, and such connectedness will substantially reshape our world in ways we can barely imagine.

References

[1] Heintzman, N.D. A Digital Ecosystem of Diabetes Data and Technology: Services, Systems, and Tools Enabled by Wearables, Sensors, and Apps. J. Diabetes Sci. Technol. 2016, 10, 35–41.

[2] Mercer, K.; Giangregorio, L.; Schneider, E.; Chilana, P.; Li, M.; Grindrod, K. Acceptance of Commercially Available Wearable Activity Trackers Among Adults Aged Over 50 and With Chronic Illness: A Mixed Methods Evaluation. JMIR mHealth uHealth 2016, 4, e7.

[3] Chiauzzi, E.; Rodarte, C.; DasMahapatra, P. Patient-centered activity monitoring in the self-management of chronic health conditions. BMC Med. 2015, 13, 1–6.

[4 Chan, M.; Estève, D.; Fourniols, J.-Y.; Escriba, C.; Campo, E. Smart wearable systems: Current status and future challenges. Artif. Intell. Med. 2012, 56, 137–156

[5] Hazarika, P. Implementation of smart safety helmet for coal mine workers. In Proceedings of the 1st IEEE International Conference on Power Electronics, Intelligent Control and Energy Systems, Delhi, India, 4–6 July 2016; pp. 1–3.

[6] Chatterjee, A.; Aceves, A.; Dungca, R.; Flores, H.; Giddens, K. Classification of wearable computing: A survey of electronic assistive technology and future design. In Proceedings of the 2016 Second International Conference on Research in Computational Intelligence and Communication Networks (ICRCICN), Kolkata, India, 23–25 September 2016; pp. 22–27.

[7] David Kotz, Carl A. Gunter, Santosh Kumar, and Jonathan P. Weiner. 2016. Privacy and Security in Mobile Health – A Research Agenda. IEEE Computer 49, 6 (June 2016), 22–30.

[8] Mohammed H Iqbal, Abdullatif Aydin, Oliver Brunckhorst, Prokar Dasgupta, and Kamran Ahmed. 2016. A review of wearable technology in medicine. Journal of the Royal Society of Medicine 109, 10 (2016), 372–380.

[9] George Boateng, John A Batsis, Ryan Halter, and David Kotz. 2017. ActivityAware: an app for real-time daily activity level monitoring on the Amulet wrist-worn device. In IEEE International Conference on Pervasive Computing and Communications Workshops (PerCom Workshops). IEEE, 431–435

[10] Andreas Lymberis and Andre Dittmar. 2007. Advanced wearable health systems and applications. IEEE Engineering in Medicine and Biology Magazine 26, 3 (2007), 29.

[12] Tracy L Mitzner, Julie B Boron, Cara Bailey Fausset, Anne E Adams, Neil Charness, Sara J Czaja, Katinka Dijkstra, Arthur D Fisk, Wendy A Rogers, and Joseph Sharit. 2010. Older adults talk technology: Technology usage and attitudes. Computers in human behavior 26, 6 (2010), 1710–1721.

[13] Vivian Genaro Motti and Kelly Caine. Human factors considerations in the design of wearable devices. In Proceedings of the Human Factors and Ergonomics Society Annual Meeting, Vol. 58. SAGE Publications Sage CA: Los Angeles, CA, 2014, 1820–1824.

[14] Alexandros Pantelopoulos and Nikolaos G Bourbakis. A survey on wearable sensor-based systems for health monitoring and prognosis. IEEE Transactions on Systems, Man, and Cybernetics, Part C (Applications and Reviews) 40, 1 (2010), 1–12.

[15] Abbasi, M. A. B., S. S. Nikolaou, M. A. Antoniades, M. N. Stevanovi´c, and P. Vryonides, "Compact EBG-backed planar monopole for BAN wearable applications," IEEE Transactions on Antennas and Propagation, Vol. 65, No. 2, 453–463, Feb. 2017.

[16] BBC (2016) Walk with the world's biggest dinosaur in virtual reality. At http://www.bbc.com/earth/story/20160219- attenborough-and-the-giant-dinosaur-virtual-reality-360, 25 June 2016

[17] Bower M, Affordance analysis–matching learning tasks with learning technologies. Educational Media International, 45(1), 3-15, 2010.

[18] Bower M and Sturman D, What are the educational affordances of wearable technologies? Computers & Education, 88, 343-353 Canberra.

[19] UC workshop: using Google Glass in class. At http://www.canberra.edu.au/aboutuc/media/monitor/2014/may/9-google-glass, 25 June 2015

[20] Coffman T and Klinger MB (2015) Google Glass: Using wearable technologies to enhance teaching and learning. Paper presented at the Society for Information Technology & Teacher Education International Conference, Las Vegas, 2-6 March 2015

[21] Robin Wright and Latrina Keith, "Wearable Technology: If the Tech Fits, Wear It," Journal of Electronic Resources in Medical Libraries, 11:4 2014, pp. 204– 216. 3. Lindsey M. Baumann, "The Story of Wearable Technology: A Framing Analysis," MA Thesis, Virginia Polytechnic Institute and State University, Blacksburg, Va., 2016. 4.

[22] Thad Starner and Tom Martin, "Wearable Computing: The New Dress Code," Computer, Issue No. 06 June 2015 (Vol. 48), pp. 12–15. 5.

[23] Kent Lyons, "What Can a Dumb Watch Teach a Smartwatch? Informing the Design of Smartwatches," Proceedings of the 2015 ACM International Symposium on Wearable Computers, pp. 3–10, 2015. 6.

[24] Sema Dumanli, "Challenges of Wearable Antenna Design," presented at ARMMS Conference, Oxford, UK, 2015.

[25] Marios Iliopoulos and Nikolaos Terzopoulos, "Wearable Miniaturization: Dialog's DA14580 Bluetooth® Smart Controller and Bosch Sensors," Dialog Semiconductor, 7 March 2016.

[26] Alex Wright, Mapping the Internet of Things, 2017.

[27] ARCEP, Livre blanc – Préparer la révolution de l'Internet des objets – ARCEP | IoT, utorité de Régulation des Communications Électroniques et des Postes, Paris, 2016.

[28] AT&T INC. 2017 Annual Report, https://investors.att.com/~/media/Files/A/ATT-IR/financial-reports/annualreports/2017/complete-2017-annual-report.pdf

[29] Automobile Mag, The Big Data Boom, Automobile Magazine, 10 October 2017, http://www.automobilemag.com/news/the-big-data-boom/

[30] Automotive News, Fiat Chrysler joins autonomous driving platform from BMW / Intel / Mobileye | EETE Automotive, http://www.eenewsautomotive.com/news/fiat-chrysler-joinsautonomous-driving-platform-bmw-intel-mobileye.

[31] BEREC (2016), BEREC Report on Enabling the Internet of Things, http://berec.europa.eu/eng/document_register/subject_matter/berec/reports/5755-berec-reporton-enabling-the-internet-of-things).

[32] CISCO (2017), Cisco Visual Networking Index: Global Mobile Data Traffic Forecast Update 2016-2021, https://www.cisco.com/c/en/us/solutions/collateral/service-provider/visualnetworking-index-vni/mobile-white-paper-c11-520862.pdf.

[33] Google Cloud Platform (2017), Designing a Connected Vehicle Platform on Cloud IoT Core Solutions, Google Cloud Platform, https://cloud.google.com/solutions/designingconnected-vehicle-platform.

[34] GSMA (2017), Mobile IoT - Internet of Things, https://www.gsma.com/iot/mobile-iot-executivesummary/.

[35] IBM (2017), Connected Cars with IBM Watson IoT, https://www.ibm.com/internet-of-things/iotsolutions/iot-automotive/connected-cars/ (accessed on 03 October 2017). [50] IEEE (2015), Toward a Definition of Internet of Things (IoT).

[36] Intel (2016), Data is the New Oil in the Future of Automated Driving | Intel Newsroom, 15 November 2016.

[37] ISO (2018), ISO/IEC JTC 1/SC 41 - Internet of Things and related technologies.

[38] ITU (2012), Overview of the Internet of things. https://www.itu.int/rec/T-REC-Y.2060-201206-I

[39] NYT (2017), BMW and Volkswagen Try to Beat Apple and Google at Their Own Game - The New York Times, 22 June 2017, https://www.nytimes.com/2017/06/22/automobiles/wheels/driverless-cars-big-datavolkswagen-bmw.html.

[40] NYT (2017), The Race for Self-Driving Cars - The New York Times, 6 June 2017, https://www.nytimes.com/interactive/2016/12/14/technology/how-self-driving-carswork.html

[41] OECD (2012), Machine-To-Machine Communications: Connecting Billions of Devices, http://www.oecd.org/officialdocuments/publicdisplaydocumentpdf/?cote=DSTI/ICCP/CISP(2 011)4/FINAL&docLanguage=En

[42] WEF (2015), Industrial Internet of Things: Unleashing the Potential of Connected Products and Services, http://www3.weforum.org/docs/WEFUSA_IndustrialInternet_Report2015.pdf.

[43] United States Department of Defense (2016), DoD Policy Recommendations for The Internet of Things (IoT), https://www.hsdl.org/?abstract&did=799676 (accessed on 09 April 2018).

[44] United States Government Accountability Office (2017), Internet of Things Status and implications of an increasingly connected world, https://www.gao.gov/assets/690/684590.pdf.

[45] Wearable Computer Applications A Future Perspective, Kalpesh A. Popat, Dr. Priyanka Sharma, International Journal of Engineering and Innovative Technology (IJEIT), Volume 3, Issue 1, July 2013.

[46] http://research.omicsgroup.org/index/Wearable-computer.

[47] https://www.wired.com/2015/04/the-apple-watch/

[48] https://developers.google.com/glass/develop/

[49] Innovation in Wearable and Flexible Antennas (book), Haider Raad Khaleel, WIT Press, 2014.

[50] http://www.computerworld.com/article/3066870/wearables/

[51] http://time.com/3669927/google-glass-explorer-program-ends/

[52] http://autismglass.stanford.edu/

[53] http://www.telegraph.co.uk/news/science/science-news/

[54] Abbas Acar, Hossein Fereidooni, Tigist Abera, Amit Kumar Sikder, Markus Miettinen, Hidayet Aksu, Mauro Conti, Ahmad-Reza Sadeghi, and A Selcuk Uluagac. 2018.

[55] Peek-a-Boo: I see your smart home activities, even encrypted! arXiv preprint arXiv:1808.02741 (2018). 26 B. Celik.

[56] Samsung SmartThings add a little smartness to your things. 2018. https://www.smartthings.com/.

[57] Cedric Adjih, Emmanuel Baccelli, Eric Fleury, Gaetan Harter, Nathalie Mitton, Thomas Noel, Roger Pissard-Gibollet, Frederic Saint-Marcel, Guillaume Schreiner, Julien Vandaele, 2015.

[58] FIT IoT-LAB: A large-scale open experimental IoT testbed. In IEEE 2nd World Forum on Internet of Things (WF-IoT).

[59] Alfred V Aho, Ravi Sethi, and Jeffrey D Ullman. 1986. Compilers, Principles, Techniques. Addison Wesley.

[60] O. Alrawi, C. Lever, M. Antonakakis, and F. Monrose. 2019. SoK: Security Evaluation of Home-Based IoT Deployments. In IEEE Security and Privacy (SP).

[61] Amazon AWS IoT 2018. The Internet of Things with AWS. https://aws.amazon.com/iot/.

[62] Android API 2018. Android Sensor API Documentation. https://developer.android.com/guide/topics/sensors/sensors_overview.html.

[63] Android Monkey 2018. UI/Application Exerciser. https://developer.android.com/studio/test/monkey. [64] Android Things 2018. Android Things Official Apps. https://github.com/androidthings.

[65] Z. Berkay Celik, Patrick McDaniel, and Gang Tan. 2018. Dynamic Enforcement of Security and Safety Policy in Commodity IoT. arXiv preprint (2018).

[66] Z. Berkay Celik, Patrick McDaniel, and Gang Tan. 2018. Soteria: Automated IoT Safety and Security Analysis. In USENIX Annual Technical Conference (USENIX ATC). Boston, MA.

[67] Haotian Chi, Qiang Zeng, Xiaojiang Du, and Jiaping Yu. 2018. Cross-App Threats in Smart Homes: Categorization, Detection and Handling. arXiv preprint arXiv:1808.02125 (2018).

[68] Shauvik Roy Choudhary, Alessandra Gorla, and Alessandro Orso. 2015. Automated test input generation for Android: Are we there yet? arXiv preprint arXiv:1503.07217 (2015).

[69] Edmund M Clarke and E Allen Emerson. 1981. Design and synthesis of synchronization skeletons using branching time temporal logic. In Workshop on Logic of Programs.

[70] James Clause, Wanchun Li, and Alessandro Orso. 2007. Dytan: a Generic Dynamic Taint Analysis Framework. In ACM Software Testing and Analysis.

[71] Paul Comitz and Aaron Kersch. 2016. Aviation analytics and the Internet of Things. In Integrated Communications Navigation and Surveillance (ICNS), 2016.

[72]Eyal Ronen, Adi Shamir, Achi-Or Weingarten, and Colin O'Flynn. 2017. IoT Goes Nuclear: Creating a ZigBee Chain Reaction. In IEEE Security and Privacy (S&P).

[73] Santa Detector 2018. IFTTT. https://ifttt.com/applets/170037p-santa-detector.

[74] Edward J Schwartz, Thanassis Avgerinos, and David Brumley. 2010. All you ever wanted to know about dynamic taint analysis and forward symbolic execution (but might have been afraid to ask). In IEEE Security and privacy (S&P).

[75] M. Sharir and A. Pnueli. 1981. Two approaches to inter-procedural dataflow analysis. Computer Science Department, New York University.

[76] Vijay Sivaraman, Hassan Habibi Gharakheili, Arun Vishwanath, Roksana Boreli, and Olivier Mehani. 2015. Networklevel security and privacy control for smart-home IoT devices. In Wireless and Mobile Computing, Networking and Communications (WiMob).

[77] SmartThings. 2018. SmartThings Community Forum for Third-party Apps. https://community.smartthings.com/.

[78] ThingsWorx 2018. PTC: Industrial IoT. https://www.ptc.com/en/about. [Online; accessed 20-June-2018].

[79] Yuan Tian, Nan Zhang, Yueh-Hsun Lin, XiaoFeng Wang, Blase Ur, XianZheng Guo, and Patrick Tague. 2017. SmartAuth: User-Centered Authorization for the Internet of Things. In USENIX Security Symposium.

[80] Raja Vallée-Rai, Phong Co, Etienne Gagnon, Laurie Hendren, Patrick Lam, and Vijay Sundaresan. 1999. Soot: a Java Bytecode Optimization Framework. In Centre for Advanced Studies on Collaborative Research.

[81] Stankovic, John. "Research directions for the internet of things." Internet of Things Journal, IEEE 1.1 (2014): 3-9.

[82] Gubbi, Jayavardhana, "Internet of Things (IoT): A vision, architectural elements, and future directions." Future Generation Computer Systems 29.7 (2013): 1645-1660.

[83] "Understanding the Internet of Things (IoT) ", July 2014. [6] Dogo, E. M. et al. "Development of Feedback Mechanism for Microcontroller Based SMS Electronic Strolling Message Display Board." (2014).

[84] N. Jagan Mohan Reddy, G.Venkareshwarlu, "Wireless Electronic Display Board Using GSM Technology", International Journal of Electrical, Electronics and Data Communication, ISSN: 2320-2084 Volume-1, Issue-10, Dec-2013

[85] Yashiro, Takeshi, "An internet of things (IoT) architecture for embedded appliances." Humanitarian Technology Conference (R10-HTC), 2013 IEEE Region 10. IEEE, 2013.

CHAPTER THREE

COMPONENTS AND TECHNOLOGIES

Introduction

When it comes to the anatomy of wearable devices, designing and selecting the right electronic components and software systems is of paramount importance because each plays a vital role in the functionality of the end product.

In spite of the variety and types of wearable devices, the majority share elemental functionalities that must be implemented in the design process. For example, most modern wearables have the following:

- At least one sensor to perceive an analog quantity (typically physical, chemical, biological, or environmental).
- A conditioning circuit that filters, amplifies, and converts the perceived signal into a digital one (Analog to Digital Conversion (ADC)).
- A processing unit along with a memory and embedded system that serves as the brain power of these smart devices where all computations and processing take place.
- A connectivity unit to transmit the captured data or receive an action command to and from another layer (a mobile phone, cloud, server, etc.) using, typically, one of

- the wireless communication technologies such as WiFi and Bluetooth.
- Input and output elements for device-user interface which may include a button, gesture pad, microphone, camera for input; an LCD display, LED lights, speaker, or other motor-based actuators for output.
- An energy source, usually rechargeable, and most likely a power management system.
- A data exchange unit for programming the unit, updating the firmware, or for simple data collection.

Noticeably, the first generation of wearables has been assembeled utilizing smartphones components and technologies. For example, during the past few years, the sensors and microprocessors of wrist and head-mounted wearable devices have been drafted off smartphones parts. This would, for many, make perfect sense since using tested components with a proven success in a newly introduced technology would help manufacturers balance features, functionality, and price for an uncertain new market.

However, electronics manufacturers today are introducing new components designed specifically for wearables. Power-efficient microprocessors, and performance-bound hardware of smaller form factor are being driven by the requirements of the wearables' next generation.

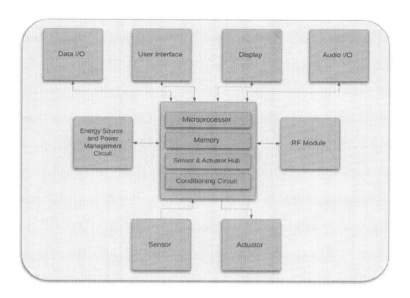

Figure 1: A block diagram of a generic wearable system

1. Hardware

1.1 Sensors

A sensor is a device that detects a change or an event in an object or environment and converts it to an electrical signal (typically). The sensor then forwards the converted signal to a computer for processing to provide a corresponding output.

Analog (continuous time, non-digital) sensors are very common in biomedical and healthcare devices. Biometric sensors such as heart rate, blood pressure, and electroencephalogram (EEG) are examples of such sensors. It should be noted that an Analog Front End (AFE) unit is needed in analog sensors, which is responsible of amplifying, filtering, conditioning the signal, and converting it to digital (using an ADC) so that it can be processed by a computer.

As of 2015, over 50% of wearable devices are based on Micro Electro Mechanical Sensor (MEMS) technology. Inertial Measurement Units (IMUs)[7] found in most of modern wrist-band wearables are perhaps the best MEMS examples. Depending on the complexity of the wearable device, it could have a single, a few MEMS sensors, or a dedicated sensor fusion interface.

The most challenging aspect in such sensors, however, lies in converting raw data into useful information. As mentioned previously, sensor fusion is the process of collecting multiple output data from multiple sensors to obtain a better total insight. A good example here is the use of data from single 3-axis acceleration along with data from a 6-axis IMU rotation sensor in fitness trackers to have more accurate information on the user's motion.

There are a number of ways to categorize sensors including the following:

    a- **Active or Passive:** Active sensors require an external power source to operate, while passive sensors simply

---

[7] ***Micro-Electro-Mechanical Systems (MEMS)*** *is a technology that is based on miniaturized electro-mechanical structures which are made using microfabrication techniques. The dimensions of MEMS devices typically range from one micron to a few millimeters. The dynamics of MEMS are controlled by the microelectronics integrated within the device.*

detect energy and operate without the need of a power source.

b- **Invasive or Non-invasive: Invasive sensors are transducers that** come into direct contact with the process (i.e.: a sensor immersed in a fluid). Non–invasive transducers, on the other hand, do not come into direct contact with the process (i.e.: an ultrasonic level sensor).

c- **Deflection or Null:** The signal produces some physical effect (i.e.: movement) closely related to the measured quantity such as in a pressure gauge where the value being measured is displayed in terms of the amount of pointer movement. In null type, which is more accurate but more complex than the deflection type, the signal produced by the sensor is counteracted to minimize the deflection. That opposing effect necessary to force a zero deflection has to be proportional to the signal of the quantity to be measured.

Sensors can also be categorized based on their area of application, mechanism, and quantities they measure.

1.1.1 Sensor Properties

A sensor should satisfy certain characteristics before integration within a system. Below are typical properties used to characterize sensors:

a- **Resolution:** The resolution of a sensor is defined as the smallest change in the input under test that can be perceived by the sensor. A high-resolution sensor is one that is able to detect a very small change in the input variable. Electronic and thermal noise in the sensor and interface circuitry can highly impact the resolution. For example, an analog temperature sensor with a resolution of 10 bits represents a range of temperature readings between 0 and 1023.

b- **Sensitivity:** It is defined as the ratio between the output signal and measured property. For example, if a temperature sensor has a voltage output, the sensitivity is then a constant with the units [v/k].

c- **Accuracy:** Accuracy can be described as the maximum difference between the actual value measured via a standard reference and the value indicated at the output of the sensor terminals. A difference value of zero indicates the highest accuracy.

d- **Precision:** It describes the reproducibility of the measurement of a given sensor and also refers to the closeness of the measurements to each other in a given scale.

e- **Drift (Stability):** Drift is a change in the sensor's reading or set point value over extended periods due to electronic aging of components or reference standards in the sensor.

f- **Hysteresis:** Ideally, a sensor should be capable of tracking the changes in the input variable regardless of which direction the change is made; hysteresis is the measure of this feature.

g- **Response Time (Responsiveness):** Response time refers to the ability of a sensor to respond to fast changes in inputs.

h- **Dynamic Range:** It refers to the full range from minimum to maximum values a sensor can measure. For example, a given temperature sensor may have a range of -40°C to +120°C.

1.1.2 MEMS Sensors

Today, the majority of IoT and wearable devices are based on Micro Electro Mechanical Sensor (MEMS) technology. Micro-Electro-Mechanical Systems (MEMS) is a technology that is based on miniaturized electro-mechanical structures which are made using microfabrication techniques. The dimensions of MEMS devices typically range from one micron to a few millimeters. The dynamics of MEMS are controlled by the microelectronics integrated within the device.

Inertial Measurement Units (IMUs)[8] found in most of modern wrist-band wearables are perhaps one of the most popular MEMS examples. Depending on the complexity of the wearable device, it could have a single, a few MEMS sensors, or a dedicated sensor fusion interface.

The most challenging aspect in such sensors, however, lies in converting raw data into useful information. As mentioned previously, sensor fusion is the process of collecting multiple output data from multiple sensors to obtain a better total insight. A good example here is the use of data from single 3-axis acceleration along with data from a 6-axis IMU rotation sensor in

---

[8] *An **Inertial Measurement Unit (IMU)** is a self-contained electronic device that measures linear and angular motion, force, and magnetic field. It is typically based on a combination of accelerometers, gyroscopes, and magnetometers.*

fitness trackers to have a more accurate information on the user's motion.

Below is a description of the most common sensors used in wearable devices:

a. Accelerometers

An accelerometer is an electromechanical device used to measure static and dynamic acceleration forces due to gravity, motion, and vibration. By measuring acceleration, the angle the device is oriented at with respect to the earth can be found, in addition to the direction of motion. In wearables, accelerometers are the most commonly used sensors. For example, an accelerometer can enable the device to distinguish between a foot step versus a wrist shake by measuring how fast a user is moving.

b. Gyroscopes

Gyroscopes are used to measure and\or maintain rotational motion and angular velocity. They are used to maintain equilibrium, determine direction and orientation of an object.

c. Magnetometers

A magnetometer is a device that measures magnetic fields. Today's digital compasses are based on magnetometers which provide the device with an orientation relative to the earth's magnetic field. A device equipped with a magnetometer will always have the Magnetic North as a reference. For example, a digital map or a display view in a smartphone rotates according on the user's physical orientation.

d. Gesture Sensors

Gesture sensors aim at enhancing the user interface with an electronic device which is usually achieved by enabling a coherent display and touchless communication. Most modern gesture sensors used in wearables utilize four directional photodiodes to sense infrared energy to convert direction, distance, and velocity information to digital information. Other gestures sensors are based on Color Sensing (RGBC), which provides red, green, blue, and clear light sensing which in turn detects light intensity under different lighting conditions. Digital Ambient Light Sensing (DALS), on the other hand, incorporates a photodiode, an amplifier, and an analog to digital convertor, in a single chip.

e. Proximity Sensors

Proximity sensors detect the presence of objects without a physical contact and produce an output in the form of an electromagnetic field or electric signal, while analyzing changes in the return signal.

These sensors use light or sound (ultra-sonic) sensitive components to detect objects, and consist of an emitter and a receiver.

f. Capacitive and Inductive Sensors

These sensors are based on a high frequency oscillator that creates a field in the immediate proximity of the sensing surface. The presence of an object in this proximity creates a change of the oscillation amplitude where the positive and negative peaks are identified by another unit that triggers the change of the sensor's output. The operation of many sensors and interface components are based on capacitive and inductive sensors, including: position, humidity, fluid level, and acceleration sensors, in addition to trackpads and touchscreens.

g. Altimeters

Most altitude sensors can determine altitudes based on the atmospheric pressure. Besides determining the user's altitude, altimeters give rise to higher processing accuracies when implemented in fitness trackers. For example, the altimeter enables fitness trackers to determine if the user is climbing stairs through sensing the height changes, which allows for a more realistic calorie loss calculation.

h. Electroencephalograph (EEG) Sensors

The EEG sensor is essentially a signal amplifier for detecting the brain's electrical activity from the head's surface where the neurons generate extremely small amplitudes of voltage.

i. Optical Heart Rate Sensors

Based on the pulse oximetry technique, these optical sensors distinguish between the optical features of the oxygenated and de-oxygenated hemoglobin. The monitor consists of a red LED and an optical detector which measures the light reflectance or absorbance during the oxygenation and de-oxygenation cycle, and the heart rate is then determined.

j. Flex Sensors

Flex sensors are passive resistive devices that changes resistance when bent which can be used to detect flexing or bending of an object.

k. Galvanic Skin Response (GSR) Sensors

Galvanic skin response (GSR) measures the continuous variation of electrical impedance of human skin which enables the detection of psychological, emotional, and physiological parameters.

l. Temperature Sensors

Low power temperature sensors drive an electric signal (voltage) that is proportional to the ambient temperature. Most small factor temperature sensors are based on thermocouples, temperature dependent resistors, often called thermistors, or temperature dependent transistors.

m. Biochemical Sensors

Integrating biochemical sensors within wearable technology is the focus of many companies who are looking for innovative ways to capture and analyze new health-related data.

Research has shown that monitoring the sodium concentration in human sweat serves as an indicator of the person's electrolytic balance and general wellbeing. To that point, recent studies report the potentials of epidermal transfer tattoo-based potentiometric sensor attached to a compact wireless transmitter for non-invasive sweat monitoring. Another example, Levl, a tech company, is working on developing a fitness tracker that features a nano breath sensor. The sensor measures the acetone concentration in breath which is released as a metabolism byproduct to determine if the user is burning fat.

Another application of biochemical sensors is based on flexible screen-printed electrodes (SPE). This technology is employed in glucose biosensors, which are used to implement wearable diabetes monitors.

n. Passive Infrared (PIR)

A passive infrared sensor is used to measure infrared light radiating from objects. They are commonly used in security alarms and auto lighting systems.

PIRs are made of a pyroelectric detector, which is an infrared-sensitive element. The sensor in motion detectors is wires as two halves to avoid measuring average IR levels. For a static object emitting IR, the produced voltages from the two halves cancel each other out. If one half detects more or less IR radiation than the other, a voltage will be produced to indicate a motion.

o. LiDAR

LiDAR, also known as laser altimetry, is an acronym for light detection and ranging. It refers to a remote sensing technology that emits a focused light wave and calculates the time it takes for the reflected wave to be detected by the sensor in order to find ranges or distances. In theory, LiDAR is similar to the old radar (radio detecting and ranging) technology, except that it is based on discrete pulsing of laser. The object's coordinates are obtained from the time difference between the transmitted laser and the received, the angle at which the laser was transmitted, and the reference location of the sensor.

1.1.3. Wireless Sensors

A crucial aspect of the sensor system is the ability to provide means of transmitting the perceived information to an external processing unit or an actuator. Obviously, a wireless transmission provides mobility, portability, and convenience, and enables the sensor to be deployed more flexibly.

Moreover, wireless sensors can be grouped together to form a network in order to provide a more sophisticated set of data and/or to communicate and exchange information. Sensors in a common network (Wireless Sensor Network (WSN)) share data either through nodes that combine information at a gateway, or where each sensor connects directly to gateways which act as bridges that connect the sensors to the internet.

1.1.4. Multi-Sensor Modules

Multi-sensor modules combine a wide range of sensors in addition to a limited-power processing unit, communication capability, cloud connectivity and other peripherals. These modules are typically used as development platforms for the design and prototyping of IoT systems.

The Texas Instruments (TI) CC2650 SensorTag powered by a single coin cell battery, for example, is one of the most commonly used modules and features the following components in a single package:

Sensor input
Ambient light sensor (TI Light Sensor OPT3001)
Infrared temperature sensor (TI Thermopile infrared TMP007)
Ambient temperature sensor (TI light sensor OPT3001)
Accelerometer (Invensense MPU-9250)
Gyroscope (Invensense MPU-9250)
Magnetometer (Bosch SensorTec BMP280)
Altimeter/Pressure sensor (Bosch SensorTec BMP280)
Humidity sensor (TI HDC1000)
MEMS microphone (Knowles SPH0641LU4H)
Magnetic sensor (Bosch SensorTec BMP280)
2 Push-button GPIOs
Reed relay (Meder MK24)

Output components
Buzzer/speaker
2 LEDs

Communications
Bluetooth Low Energy (Bluetooth Smart)
ZigBee
6LoWPAN

The module uses a processing module that includes an extremely low power CPU (ARM Cortex M3) with a 128 KB of flash memory and 20 KB of Static RAM (SRAM). While power-efficient, this limits the amount of processing and resources on this system. Typically, such limited-power devices will need to be supplemented by a gateway, router, smartphone, etc.

Another module used for IoT and wearable devices prototyping is the PRISM introduced by Eleco with the following specification:

Communications: Bluetooth 4.0 (BLE)
Microcontroller: ARM Cortex-M0+ 32 bit
Power source voltage: +2.35 to 3.3 V
Consumption current: 5 mA (Peak current)
Standby current: 8 μA
Accelerometer: 3-axis ±2 G (max. ±16G)
Compass: 3-axis ±1300 μT
Thermometer: -40 to +120°C
Hygrometer (A hygrometer is a sensor used to measure humidity and water vapor in the atmosphere, in soil, or in confined spaces): 0 to 100%
Barometer: 300 to 1200 hPa
Illuminometer: 0 to 128 kLx
UV meter: UV index 0 to 11+

1.1.5. Signal Conditioning for Sensors

The voltages produced by analog sensors are extremely small (ranges from pico-volts to milli-volts). Thus, they need to be amplified before they can be used as an input to the analog to digital conversion stage. Such amplification is only one part of the signal conditioning process. Impedance matching, input-output isolation, and filtering may also be required before the signal can be processed and analyzed.

1.2. Actuators

Unlike sensors that provide information about a process/environment, actuators provide action. Actuators typically receive some type of control signal based on sensors' data that triggers a physical effect.

Actuators can also vary in type, function, and area of application. Some common categorizations are based on power, motion, and industry.

The most powerful use cases in IoT and wearable technology are those where sensors and actuators work together in an intelligent, complementing, and harmonious fashion. Such combination can be utilized to solve problems by simply elevating the data that sensors provide to actionable insight that can be acted on by work-producing actuators[9]. Examples of actuators include motors, relays[10], speakers, and lights. Just like a sensor, an actuator may need a conditioning and/or driving circuit. Fig. 2 depicts a signal flow in a sensor/actuator based system.

1.3. Microcontrollers, Microprocessors, SoC, and Development Boards

The most essential component and what makes IoT and wearables a smart technology is either a microprocessor or a microcontroller.

---

[9] *Haptics*, *also known as kinaesthetic communication refers to the recreation of touch experience by applying forces, vibrations, or motions to the user using a variety of actuators.*

[10] *A relay is a binary actuator that has two stable states, either latched (when energized), or unlatched when de-energized. The most popular relays are: electromagnetic relays, which are constructed with electrical, mechanical and magnetic components, and have operating coil and mechanical contacts; solid state relays, which use solid state components to perform the switching mechanism without moving any parts; and hybrid relays.*

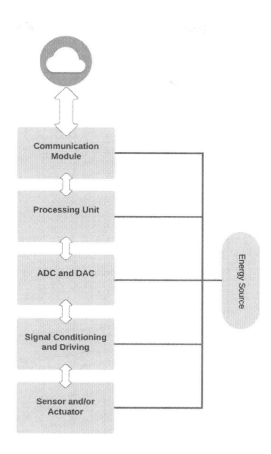

Figure 2: Sensor and actuator signal flow in a system

A microprocessor incorporates the functions of a computer's central processing unit (CPU) in a single integrated circuit (IC) chip. It accepts digital data as input, processes it according to a sequence of instructions stored in the memory, and delivers results as output. Microprocessors are used in everything from the smallest handheld devices to the most powerful supercomputers. In addition to the CPU, microcontrollers also have a RAM, ROM and other peripherals integrated within a single chip.

The choice of the microprocessor in IoT and wearables is driven by the application, industry, and functions performed by the device. For most applications, a general purpose microprocessor unit would suffice; however, highly specialized devices would most likely require a dedicated application processor. It should be noted that nowadays microprocessor manufacturers incorporate most of the functions in a single chip which is crucial in reducing the overall size and cost of a wearable device. For example, the 32-bit ARM processor, which is a reduced instruction set computer (RISC) architecture developed by Advanced RISC Machines (ARM) is very common in IoT and wearables as it provides a sufficiently powerful computational performance and energy efficiency.

Another popular microcontroller is the Programmable System on Chip (PSoC)[11] developed by Cypress Semiconductor, integrates programmable analog and digital functionalities in a single chip utilizing the power of an ARM cortex-M core architecture.

It is also worth mentioning that some advanced devices have a separate co-processor (sensor hub) dedicated to handle the sensors' data. This is critical when the device has a large amount

---

[11] *A **System on Chip (SoC)** refers to a grouping of all the components of an electronic system in a single-integrated circuit. In addition to the processing unit, memory, bus, a SoC may contain a sensor(s), communication capability, and other components that deal with data compression, data filtering, etc.*

of sensors data that needs to be processed in real time, which requires an uninterrupted CPU attention. This function is widely known as 'sensor fusion'.

It should also be noted that based on the features offered by the IoT or wearable product, the device may or may not require a specific operating system. For instance, a light-weight RTOS (Real Time Operating System) may be more than sufficient to operate a wristwatch that measures temperature, tracks a user's movement using a simple accelerometer, and displays time on a basic LCD display. On the other hand, a sophisticated smart watch that serves as an extension of a user's mobile phone needs to run an advanced operating system such as an iOS or Android.

A development board, on the other hand, is a prototyping solution that features a low-power CPU which typically supports various programming environments. The board, in essence, is a printed circuit board containing a microcontroller unit, interfacing circuitry, power management unit, and communication capability. A supporting firmware in addition to data transfer to a cloud-based server is typically included.

Developing IoT and wearable applications is more accessible with the growing availability of low-cost, off-the-shelf development boards, platforms, and prototyping kits. Such modular hardware offers greater flexibility to the designer.

While a microcontroller is a SoC that provides data processing and storage capabilities, a single board computer (SBC) is a step up from microcontrollers. They allow the user to connect peripheral devices like keyboards, and screens, in addition to offering more processing power and memory.

Sensors and actuators connect to the microcontroller and microcomputer through analog and digital General Purpose Input/Output (GPIO) pins or through a bus. Standard communication protocols like SPI and I2C are used for in-device communication.

The ARM Cortex-M processors are very widely used in IoT and wearables. They support multiple clock and power domains. They also support advanced low power techniques and provide different sleep modes. Below is a comparison between different M processors:

Cortex-M0 processor: The smallest ARM processor which makes it ideal for low-cost microcontrollers for general data handling and simple input-output control tasks.

Cortex-M0+ processor: The most energy efficient ARM processor. It features the same instruction set as Cortex-M0, in addition to IoT and wearable application it is suitable for general data handling, and input-output operations.

Cortex-M3 processor: This one is by far the most popular ARM processor used in high performance microcontrollers that are also energy efficient.

Cortex-M4 processor: This processor features all the functions of the Cortex-M3 processor, with additional instructions to support DSP operations.

Cortex-M7 processor: The highest performance processor with open memory interface options and a more powerful DSP performance.

Cortex-M23 processor: This processor is similar to the Cortex-M0+ processor, but features a newer architecture version called ARMv8-M, which adds a security extension and several additional instructions.

Cortex-M33 processor: This processor is similar to the Cortex-M3/M4 processor but also supports ARMv8-M architecture in addition to enhanced system level features.

FPGA (Field-Programmable Gate Arrays) are integrated circuits, which are groups of programmable logic gates, memory, and other elements. FPGAs could be coupled with a processor to interface with the outside world very easily and provide lowest power, lowest latency and best determinism, and to leverage more advanced software functions such as web services or security packages. Fig.3 shows a typical FPGA development board.

Figure 3: The Arm MPS2+ FPGA Prototyping Board. The platform offers a large FPGA for prototyping Cortex-M based designs with a range of debug options (Photo courtesy of ARM)

1.3.1 Selecting the Right Processing Unit for Your Wearable Device

There are a plethora of development boards, microcontrollers, and microprocessors available in the market, and selecting the right one is dependent on a number of factors and is highly bounded by the targeted application. Below are some important factors the product developer should be aware of:

- **Compatibility:** Does the unit support the sensors and actuators required in your project?

- **I/O Support:** The number of Input/Output ports determines the number of sensors and actuators to be used in the project.

- **Architecture**: Can the architecture handle the complexity of your project? Selecting the right architecture depends on the functional requirements of your project and how much computing power your application will need.

- **Clock Speed and Memory:** Is the processing unit equipped with adequate memory for your project? Also, some IoT or wearable applications will run adequately at low speeds, some will run the processor at higher speed to achieve a more demanding task, some may have different clock needs depending on the dynamics of the application. The designer needs to make an informed decision concerning this before prototyping.

- **Power Requirement:** How much power will the unit need? What is its power consumption while in action and during idle time? Energy efficiency is extremely important for wearables and mobile/portable (non-wired) IoT applications.

- **Customer and Community Support:** Is good documentation for your unit available? Is customer support reputable and reliable? This is crucial when it comes to making informed decisions on how to professionally use your unit.

- **Add-On Capabilities:** Some wearable and IoT applications may require Digital Signal Processing (DSP) capability for analysis and modification of signals. Hence, a dedicated digital signal processor may be required onboard.

- **Connectivity:** Most wearables must have a form of connectivity. The availability of connectivity type(s) such as Ethernet, WLAN, and BLE, needed for the project on board is a great advantage.

- **Security:** Security is of paramount importance in wearable devices. Hardware support for security may be required or preferred in some applications.

Because IoT and Wearable Technology cover a wide spectrum of applications, processing and wireless requirements vary drastically. For example, some wearable devices perform a small amount of processing and merely uploading data to the cloud. Such devices use low-cost, low-power microcontrollers. The wireless connectivity is typically integrated within the board. Other devices, such as smart watches and security cameras, require upper-scale processors for data analysis or driving a display.

It is also worth noting that many smart devices in the market today use repurposed smartphone processors. However, other companies have gone the extra mile and designed processors dedicated for IoT and wearable devices.

1.4 Wireless Connectivity Unit

Clearly, wireless connectivity is of paramount importance in wearable devices as most of them need to interact with a networking device. Supporting one or more wireless communication protocols such as Wi-Fi, Bluetooth Low Energy (BLE), and IEEE 802.15.4 LR-WPAN (Low Rate Wireless Personal Area Network) is typically required in these devices.

Obviously, no wireless transmission is possible without an antenna, and thus, the functionality and efficiency of any wearable device with a wireless connectivity is primarily dependent on the properties of the integrated antenna unit[12].

---

[12] *Antennas are electromagnetic radiators that convert electrical currents to electromagnetic waves at the transmitting end and from electromagnetic back to electrical currents on the receiving end. The most common types of antennas in terms of radiation are omni-directional and directional antennas. Omni-directional antennas radiate its energy in all directions equally except top and bottom (donut shaped radiation pattern) whereas a directional antenna will focus its energy in a certain direction. Common antennas in IoT and wearable devices are: chip, PCB, and wire antennas.*

In general, the small form factor nature of wearable devices requires the integrated wireless connectivity components to be compact, light-weight, low-profile, and mechanically robust, simultaneously. They also must exhibit reliability, high efficiency, and desirable radiation characteristics.

In wearable technology, there is a number of additional challenges that engineers face when designing antennas and wireless systems that do not exist in conventional wireless units which will be discussed in Chapter 4.

Designers must choose between laying out their own Radio Frequency (RF) transceiver chip, antenna, and impedance matching circuit in the form of a chip or a PCB, or going for off the shelf RF modules. Designing a dedicated PCB that meets electromagnetic compatibility (EMC), electromagnetic interference (EMI), and regulations could be a very lengthy and expensive process.

The commercially available system-in-package (SiP) RF modules integrate all the necessary components including the antenna, and they typically come pre-tested and pre-certified. That minimizes a lot of design complexity and reduces development time, energy, and risks, which allow developers to focus on their target applications. Figs. 4 and 5 show a printed monopole antenna and typical radiation patterns of monopole/dipole, and microstrip antennas, respectively.

Figure 4: Printed monopole antenna intended for integration within flexible electronics

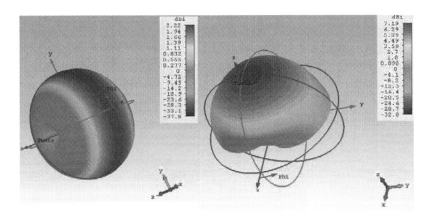

Figure 5: Omni directional radiation pattern (left), and Semi-directional (hemi-spherical radiation pattern (right)

1.5 Battery Technology

When designing a wearable product, it is important to consider throughout the design process how the performance of the device will affect its energy budget. Energy consumption, battery capacity, and duty cycles are among the key components of the energy budget.

Although computer architects are striving to produce ultra-low power microprocessors and microcontrollers, the power demand is still high due to a larger size, higher resolution displays, and multiple apps that are in use simultaneously. Unfortunately, there is no version of Moore's law that applies to batteries as the annual improvement rate in battery capacity does not exceed 8%. According to a recent study, one-third of Americans who have a wearable device stop using it within the first six months due to battery life limitation[13].

Lithium-based rechargeable batteries have become the obvious choice in hand-held applications since its commercial debut in the early 1990's. There are several reasons attributed to its dominance: higher cell voltage compared to Ni-based batteries, lighter weight, higher energy density, relatively simpler manufacturing process, and higher recharge-ability rate. It should be noted that the principal difference between Lithium ion and Lithium polymer (the main Li-based battery technologies) is that

---

[13] ***Moore's Law*** *refers to an observation pointed out by Gordon Moore (a co-founder of Intel Corp.). His observation concludes that the number of transistors per square inch in integrated circuits doubled each year. Although the rate seemed to hold true from 1975 until around 2012, the rate started to slow in 2013, and in 2015 Gordon Moore himself stated that the growth rate would reach saturation in the following decade.*

Lithium Ion has a higher capacity whereas the Lithium Polymer is lighter in weight.

On the other hand, both flexible and printed batteries offer promising compact solutions for wearable devices such as in transdermal drug delivery patches, temperature sensors, and RFID tags. These include polymeric lithium, solid-state, printed zinc-based batteries, in addition to flexible supercapacitors.

Another compact option is using a miniaturized packaging of traditional batteries, as in the battery offered by Panasonic that is available in a cylindrical package of 3.5 mm diameter and 2 cm length.

It is also worth noting that wireless battery charging is already available in the market for smartphones and related accessories, and will be naturally adopted by IoT and wearable devices. Wireless charging provides efficient power transfer to batteries through either RF power over a distance or via inductive coupling where a transmitting coil provides an electromagnetic field that transfers energy to a closely positioned receiving coil.

In some applications, the relatively short battery life adds significant cost to a given system over its lifetime. The battery unit itself may be inexpensive, but the costs associated with the system downtime when the battery is drained (or during replacement) can add additional costs over a product's lifecycle.

Solar energy is potentially capable of harvesting significantly more power than many other alternative energy sources like: thermoelectric and piezoelectric transducers, electrodynamic switches, and ambient RF signals. Hence, integrating a solar panel into IoT, and in some applications, wearables, could be a tangible solution. Solar panels can harvest light energy both indoors and outdoors providing a source of consistent power, thus increasing the lifetime of the product while reducing the total cost to the end user.

1.5.1 Power Management Circuits

Embedding signal conditioning within the sensor provides some considerable advantages. The data that is sent to the microcontroller unit will be swiftly and easily interpreted by the application, which gives rise to less power being consumed by the microcontroller.

Depending on the type of battery used in the device, there is often a requirement for step-up boost converters or boost-switching regulators[14]. A careful choice can make a huge impact on the system's overall power consumption.

For more complex devices, a power management integrated circuit (PMIC) provides a more precise control over the entire system. From a single power source, one can derive multiple voltage rails to power different components of an embedded system. A PMIC may also offer additional functionality for general system control, such as timers, voltage sequencing, and reset capability.

It is also worth noting that in addition to using low-power semiconductor components, utilizing software techniques, including stacks, encryption and data processing, are key

---

[14] ***Step-up Switching Converters***, *also called boost switching regulators, enable a higher voltage output than the input voltage. The output is regulated, as long as the power draw is within the specified output power.*

considerations. Each of these design factors can have a significant impact on the system's overall power budget.

1.6 Displays and Other User Interface Elements

How a user interacts with a wearable device is an essential design aspect. Complexity should be minimized, and the interactive experience should be as intuitive as possible.

Inherited from mobile phones and hand-held electronics, displays with capacitive touch screen capability are the obvious choice in providing a user with a feedback and/or a user interface platform.

The emergence of flexible and curved displays is considered as the major breakthrough in wearable display technology, and their introduction to the electronics market was certainly driven by wearables.

Organic Light Emitting Diode (OLED) based displays, which emits light from an organic compound layer in response to electric current, has substantial performance advantages in wearables. By only powering the active pixels, a considerable amount of energy can be saved. Furthermore, OLED offers a borderless and semi-transparent projection which gives rise to an improved user experience especially in smart glasses applications.

Another noteworthy example of wearable technology-driven development in display technology is the Digital Light Processing (DLP). It has many performance gains over traditional Liquid Crystal Display (LCD)-based projectors since it enables noise-free, precise image quality, and color reproduction capability.

In addition to touch screens, buttons, switches, knobs, and sliders there are other ways a user can electronically interact with the smart device. Other important elements used in IoT and wearables include buzzers and vibrating motors which are essential in alerting the user when certain activities take place. For instance, the vibrating motor in smart watches is utilized to alert a user when a message is received. LEDs and digital segment displays are also very commonly used in wearables to provide feedback.

1.7 Microphones and Speakers

Many wearables have integrated microphone and speaker to perform voice commands through a user interface platform. Different types of microphones and speakers can be embedded in these devices including piezoelectric MEMS microphones and micro-speakers.

2. Architecture, Software, and Communication Technologies

Complexity is one of the biggest challenges that face the designer when planning a wearable solution. A characteristic solution involves a number of heterogeneous wearable devices, with sensors that generate data which is then analyzed to provide insights. Further, a myriad of wearable devices is connected through a gateway device to a network. The job of a gateway is to enable the devices to communicate with each other and with cloud services and applications. Thus, we need to develop a process flow for a concrete framework over which a wearable solution is built.

In general, the architecture portrays the structure of IoT and wearable solutions including the physical aspects (i.e.: devices, sensors, actuators, etc.), and the virtual aspects (i.e.: services, protocols). It is worth noting that there is no single IoT architecture that is agreed upon universally by the technical communities. Various architectures have been proposed by different researchers and technical bodies. However, adopting a multi-layered architecture allows the designer to focus on improving the understanding about how all of the aspects of the architecture operate independently before they are integrated into an application. Such modular approach supports managing the complexity of IoT and wearable solutions.

As most designers know, even the simplest project requires careful planning and an architecture that comply with a set of standards. Furthermore, when projects become more complex, detailed architectural plans are often required by law. For data-driven IoT applications, a basic three-tiered architecture can be used to understand the flow of information from smart devices, through a networking element(s), and out to the cloud services. A more elaborate IoT architecture would include additional vertical layers that cut across the other layers, such as data management and information security.

2.1. IoT Architectures

The new challenges and requirements of IoT are driving an entirely new area of network architecture. In the past decade, architectural standards and frameworks have materialized to address the challenge of deploying extremely large-scale IoT networks. The underlying concept in all these architectures is to support data, processes, and functions that IoT devices would perform. Common IoT architectures include the Basic Three-Layer IoT Architecture, the oneM2M Architecture, and the IoT World Forum (IoTWF) Architecture, etc. The reader is referred to [35] for a more technical and in-depth information on these architectures.

2.2 Wearable Device Architecture

One could utilize one of the IoT architectures mentioned in the previous section to design sophisticated wearable devices. It should be noted, however, that due to the size constraints, and limited computational power of wearables, a typical system architecture would include a gateway to the Internet (i.e.: a smart phone) with a dedicated application used for configuring the device and for processing the perceived data. The gateway device is also used for data visualization purposes where relevant processed data is translated into a graphical representation. The perceived data is forwarded wirelessly using a low power transceiver (i.e.: Bluetooth Low Energy (BLE)) for basic processing. A more thorough and insightful data processing takes place in the cloud. Processed data is then transferred back to the wearable device for feedback while a copy of it is archived at the data center. Obviously, the gateway provides the means of network connectivity for such back and forth communication. An architecture for a typical wearable device is depicted in Fig. 6.

It should be noted that wearables can be either simpler or more complex than what's shown in Fig. 6. For instance, some wearable navigators used by professional hikers connect directly to a Global Positioning System (GPS), thus bypassing the gateway layer. On the other hand, some early fitness trackers relied only on a smart phone for processing and feedback without the need to use a cloud service.

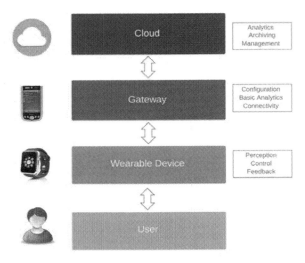

Figure 6: A basic architecture for wearable devices

2.3. Operating Systems

Based on the features offered by a wearable product, the device may or may not require a specific operating system. For instance, a light-weight RTOS[15] (Real Time Operating System) may be more than sufficient to operate a wristwatch that measures temperature, tracks a user's movement using a simple accelerometer, and displays time on a basic LCD display. On the other hand, a sophisticated smart watch that serves as an extension

---

[15] *A **real-time operating system (RTOS)** is an operating system aimed to serve real-time applications which processes information as it arrives in, without any buffering delays.*

of a user's mobile phone needs to run an advanced operating system such as an iOS or Android.

2.4 Communication Protocols and Technologies

Most wearable devices must connect to a network for their data to be utilized. In addition to the wide range of components that make up these devices, there are also several communication technologies and protocols used to connect them.

Protocols ensure that data from one device or sensor is reliably and securely delivered and understood by another device or system. Given the diverse types of wearable devices available, using the right protocol in the right context is of paramount importance.

There exists an overwhelming number of connectivity options for designers working on products and systems for IoT and wearable technology. How protocols and standards support secure and reliable data exchange in the ecosystem is a question that any experienced designer should know the answer to. It is important to take the application requirements, architecture, and factors that impact signal quality, interference, bandwidth, and range into account.

The physical and data link layers comprise devices and physical networks connecting them with other devices, network, and/or gateways. When designing a new connected product, there is a bewildering number of protocols, standards, and technologies to choose from. In a perfect world, networks would consume extremely small amount of power, offer a very wide range of coverage, and have a very large bandwidth. Unfortunately, this does not exist at the moment.

The available connectivity options are all governed by a tradeoff between power consumption, range, and bandwidth (data rate). Without focusing on wired technologies[16], below are some of the popular physical and data link layer wireless protocols and technologies:

2.4.1 Short Range

2.4.1.1 Bluetooth (Short Range, High Data Rate, Low Power)

Bluetooth is a short-range wireless communications technology standard that can be found in most smartphones and portable devices, which offers a major advantage for wearables and other personal gadgets.

Bluetooth is well-known technology for a long time, but not long ago a new WPAN technology introduced by the Bluetooth Special Interest Group (Bluetooth SIG) aimed at novel applications in the healthcare, wearables, security, and home entertainment industries. Compared to the legacy Bluetooth, Bluetooth Low Energy (BLE), formerly known as Bluetooth Smart, provides considerably reduced power consumption and cost while maintaining a comparable range.

Bluetooth can run various applications over different protocol stacks, however, each one of these stacks uses the same Bluetooth link and physical layers.

---

[16] *Ethernet for IoT is a simple and inexpensive wired connection solution that provides fast data connection and low latency in stationary IoT applications.*

2.4.1.2 NFC and RFID (Short Range, Low Data Rate, Low Power)

NFC (Near Field Communication) offers a set of communication protocols and technologies using electromagnetic fields that enable simple and secure two-way interactions between electronic devices.

NFC has its origins in Radio Frequency IDentification (RFID) technology, which uses electromagnetic radiation to encode and receive information. Any NFC-enabled device has a microchip that is activated when it comes in close proximity to another NFC-enabled device (typically less than 10 centimeters).

NFC solves many of the challenges associated with IoT and wearable devices such as offering a simple tap-and-go mechanism which makes it easy and intuitive to connect two different devices. Also, the short range of NFC prevents against unauthorized access.

2.4.1.3 Z-Wave (Short Range, Low Data Rate, Low Power)

Z-Wave is a low-power wireless communications technology that is primarily designed for IoT products such as smart lighting, smart locks, and security and alarm among many others.

This sub-1GHz band technology is designed for reliable and low-latency communication of small data packets. It is scalable (supports up to 232 devices), and supports mesh networks without the need for a coordinator node.

2.4.2 Medium Range

2.4.2.1 Wi-Fi (Medium Range, High Data Rate, High Power)

The same good old technology that connects most of our computers and gadgets to the internet can be used to connect IoT and wearable devices as well. Because Wi-Fi consumes a relatively higher energy compared to other technologies, it's often overlooked for battery-operated devices, but its pervasiveness and low cost makes it a viable option for certain applications. Wi-Fi, depending on the operating frequency (2.4GHz and 5GHz bands), and number of antennas can offer different ranges (up to 70 meters indoor), and data rates (600 Mbps maximum, but 150-200Mbps is more typical).

2.4.2.2 ZigBee (Medium Range, Low Data Rates, Low Power)

ZigBee-based networks are characterized by low power consumption, low data rates (up to 250 kbps) and a line of sight connectivity range of up to 300 meters and 100 meters for indoors.

The ZigBee standard is a relatively simple, easy to install, scalable to thousands of nodes, resistant to communication errors and unauthorized readings, and has high security and robustness. Typical applications include wireless sensor networks (WSNs) in M2M, IoT, and wearable technology applications.

2.4.3 Long Range

2.4.3.1 LPWAN and LoRa (Long Range, Low Data Rate, Low Power)

Low Power Wide Area Network (LPWAN) is a type of network that serves the needs of applications requiring long distance communications but also with limited power budget.

LoRa (Long Range) is an LPWAN technology that uses license-free sub-gigahertz bands like 433 MHz, 915 MHz, and 923 MHz. LoRa enables long-range (2-5 km in urban areas, 15 km in suburban areas) with low power consumption.

2.4.3.2 Sigfox (Long Range, Low Data Rate, Low Power)

Sigfox is an ultra-narrowband (UNB) technology based on binary phase-shift keying (BPSK). Sigfox encodes the data by taking very narrow slices of spectrum and changes the phase of the RF carrier signal. This allows the receiver to only tune in to a small slice of spectrum aiming at mitigating the effect of noise. Achievable data rate in Sigfox is modest (up to 1 kbps) but it can support a wide range of up to 50 km in open areas with a very low power consumption.

2.4.3.3 Cellular Technology (Long Range, High Data Rate, High Power)

Cellular technology is the basis of mobile phone networks but it could also serve as a platform for IoT applications that require long distance communication. Cellular technology is capable of transferring large amounts of data but at the expense of high power consumption and cost.

Global System for Mobile Communications (GSM) has also been used for IoT systems represented by Extended Coverage GSM IoT (EC-GSM-IoT) which is a standard-based Low Power Wide Area Network technology. It is based on e- General Packet Radio Services (eGPRS) and implemented as a long-range, high capacity, and low energy cellular system for IoT communications. 4G Long Term Evolution (LTE) networks also support IoT. However, the exponentially growth in IoT market has kept LTE networks struggling to keep up with the resource demands. The solution is: 5G. As described by the International Telecommunication Union (ITU), all usage scenarios for 5G networks support IoT devices: massive Machine-Type Communications (mMTC), enhanced Mobile Broadband (eMBB), and Ultra-Reliable and Low-Latency Communications (URLLC).

2.5. Cloud

After your wearable technology project is up and running, devices will start to generate vast amounts of data. An efficient, scalable, and cost-effective means will be needed for managing those devices and handling all that information and deliver the desired outcomes for you. When it comes to long-term storage, processing, and data analysis, there's nothing can beat the cloud.

By minimizing the need for on-premises infrastructure, the cloud has enabled businesses to go beyond the conventional applications of wearable devices, and accelerated the large-scale deployment of these technologies. Moreover, as data from the physical world comes in various formats, cloud platforms offer a wide range of management solutions from unstructured bits of data, such as images or videos, to structured entities, and high-performance databases for telemetry data.

On the other hand, Edge computing where data is processed closer to the endpoints is increasingly being employed in IoT and wearable technology to cut down the latency and expedite the decision making process. Current deployments often employ a mix of cloud and Edge computing to get the best of the two worlds. For example, health monitors and other healthcare wearable devices can save lives by instantaneously alerting medical staff when help is needed. Moreover, smart surgical assistive devices must be able to analyze data swiftly, safely, and accurately. If these devices strictly rely on transmitting data to the cloud for decision making, the results could be disastrous.

2.5.1. Why Cloud?

IoT and wearable technology cloud comprises the services and standards necessary for connecting, managing, and securing a wide spectrum of devices and applications enabled by these technologies, in addition to the underlying infrastructure required for processing and storing the data produced by these devices. The cloud enables businesses to leverage the potential of these technologies without having to build the necessary infrastructure and services from the ground up.

The cloud offers a more efficient, scalable, and flexible model for bringing the infrastructure and services to power IoT and wearable devices and their applications. Most of IoT is virtually limitless in scale, unlike most organizations' resources. The cloud computing model effectively takes in the ever-expanding scale of IoT and wearable devices, and it can do so in a cost-effective manner.

2.5.2. Platforms

A cloud platform for IoT and wearable technology is an essential component of their massive ecosystem. Since not all wearable devices need cloud services, we typically refer to such platforms as IoT platforms as an umbrella term.

An IoT platform is a multi-layer technology that facilitates provisioning, automation, and management of connected devices. It essentially connects a diversity of hardware to the cloud utilizing enterprise-grade security mechanisms, data processing capabilities, and connectivity options. An IoT platform provides a set of ready-to-use features for developers that could considerably speed up the development of applications for IoT and wearable devices and cut down significant costs. Moreover, platforms are perfect when it comes to scalability and device heterogeneity.

Initially, IoT platforms were intended to act as a middleware, i.e.: to function as a mediator between the hardware and application layers. To be practical, IoT middleware is expected to support interfacing with any type of connected devices and merge in with third-party applications without any issue.

Below are some of the popular IoT platforms available in the market currently:

- **Amazon Web Services (AWS)**

    The cloud services provided by Amazon comprise an IoT suite that supports all aspects and needs of IoT applications. Examples of IoT services provided by AWS include AWS IoT Core, which deals with building IoT applications; AWS IoT Device Management which allows straightforward addition and organization of devices; AWS IoT Analytics, which provides a service for automated analytics of large amounts of diverse types of data from different types of devices; and AWS IoT Device Defender, which supports the configuration of security mechanisms for connected devices.

- **Google Cloud IoT**

  Google Cloud is another global platform that supports IoT solutions. Its IoT package enables the developers to create and manage systems regardless of size and complexity. Dedicated IoT services include: Cloud IoT Core, Cloud Pub/Sub, and Cloud Machine Learning Engine.

- **Microsoft Azure IoT Suite**

  Microsoft Azure, is another global cloud service provider in the same league as AWS and Google Cloud Platform. Azure IoT Suite offers both preconfigured and customizable solutions. Service packages similar to the ones offered by AWS and Google are available too.

Other major platforms include SAP, Salesforce IoT, Oracle Internet of Things, Cisco IoT Cloud Connect, IBM Watson Internet of Things, GE predix, Autodesk Fusion Connect, ThingWorx (now acquired by PTC), and Xively Platform.

Fog platforms also exist. For example, AWS IoT Greengrass extends AWS to edge devices so they can operate locally on the data they generate, while the cloud is still used for management, analytics, and archiving/storage purposes. With AWS IoT Greengrass, IoT and wearable devices can run AWS Lambda functions, use machine learning models, sync data in devices, and establish secure communication with other devices, even when they are not connected to the Internet.

With AWS IoT Greengrass, you can use familiar programming languages and models to create and test your device software in the cloud, and then deploy it. AWS IoT Greengrass can be programmed using familiar languages and programming models to filter device data, manage the device data, and only send necessary information back to AWS. AWS IoT Greengrass Connectors can also be used to connect to third-party applications, on-premises software, and other AWS services.

AWS IoT Greengrass lets the developers create IoT solutions that connect different types of devices with the cloud and each other. Devices that run Linux, such as Raspbian, Arm, and x86 architectures can host AWS IoT Greengrass Core which enables the local execution of AWS Lambda code, security, messaging, and data management.

2.5.2.1 Criteria for Choosing a Platform

As seen from the previous section, there are numerous IoT platform options to choose from, which makes it difficult to find the best solution for the project in hand. Below are the major criteria for choosing an IoT platform:

- **Cost and Payment Model:** Some platforms use the pay-as-you-go model where the client is charged only for the resources they actually consume (e.g.: AWS IoT Core), while other platforms use the subscription model with a flat fee monthly bill (e.g.: Salesforce). Depending on the project needs, one should choose the payment model that works best.

- **Platform Stability:** With so many platforms in the market, it's likely that some will go down at some point. It's important to choose a platform from a reputable vendor that will likely to be around for several years.

- **Platform Scalability and Flexibility:** In many cases, the project needs will change with time. Developers have to make sure that the chosen platform can accommodate the needs of the project if scaled up. In addition to scalability, the platform should be flexible enough to keep up with the newly emerging technologies, protocols, and features. Flexible platforms are often those that are built on open standards and that commit to keeping pace with the rapidly changing protocols and standards. It's also crucial that the platform is unbound to hardware and network.

- **Time to Market:** As mentioned previously, one of the greatest advantages of using an IoT platform is that it accelerates the time to market. A realistic estimate of how long the deployment process takes to get to market should be inquired prior to making the deal. Data analytics capabilities and data ownership are also important factors to consider when choosing a platform.

2.6. Data Analytics and Machine Learning

One of the core subjects in IoT and wearable technology is how to make sense of the massive amount of data that is generated. As mentioned previously, the real impact of data coming from smart devices is realized only when the analysis of the data leads to actionable business insights.

Because much of this data can seem beyond grasp, specialized algorithms and tools are needed to find the data relationships that will lead to useful insights. This brings us to the topic of machine learning.

Machine learning is part of a larger set of technologies commonly grouped under the umbrella of Artificial Intelligence (AI). Once collected data is analyzed, intelligent actions need to be taken. Performing such analysis manually is close to impossible, or very impractical.

The most useful feature of machine learning in IoT and wearable technology is that it can detect outliers and abnormal activities and trigger necessary actions accordingly. As it learns more and more about an event or activity, it gets more accurate and efficient. Moreover, machine learning algorithms can create models which predict future events precisely by identifying the factors that lead to a particular result.

The difficulty, however, lies in determining the right algorithm and the most appropriate learning model for each use case. Such analysis goes beyond the scope of this chapter, and the reader is referred to a couple of resources that can be found in the references section of this chapter ([29 and [30]).

2.6.1. Artificial Intelligence

Artificial Intelligence (AI) is a computer-based system that is capable of performing tasks that require human intelligence, such as visual perception, voice recognition, and decision-making.

The deployment of AI in wearable devices is at its infancy. An innovative approach of integrating AI into wearables was recently introduced by PIQ, a French start-up specializing in sports wearables. Their system is claimed to identify athletes' success elements by breaking down and analyzing athletic movements via specialized motion-capture algorithms and ultra-high performance sensors.

2.6.2 Machine Learning

Most scientists agree that Machine Learning (ML) is a subfield of AI and its root can be traced back to the field of Neural Networks. ML emphasizes on prediction and estimation, based on known features learned from the training data.

In the realm of wearables, Atlas fitness wearables uses ML algorithms that automatically classifies the user's exercise routine in a three-dimensional vector, being able to distinguish between regular and triangle pushups. As reported by Atlas: aside from exercise detection, the future of ML algorithms lies in the body language and habitual movements. This can give indications about the user's mood, energy level, and physical reaction by analyzing data acquired from walking, sitting, sleeping or interacting with other people.

2.6.3. Data Mining

The data mining field stems out of Knowledge Discovery in Databases (KDD), where the emphasis is on the discovery of previously unknown features of a given data.

Data mining would be valuable in wearables where a great deal of data from various sensors is available for analysis. Data mining algorithms are applied in many tasks including anomaly detection and prediction, and decision making. This could be utilized to analyze biomedical data sets and provide a synopsis of the data properties used in experimental validation.

2.7. Other Software Technologies

2.7.1. Virtual and Augmented Reality

Virtual reality (VR) refers to the use of special computer software and hardware to produce realistic images, sounds and other perceptions that replicate an actual environment or create a fictitious one. By using specialized displays or projectors and sensory peripherals the user's physical presence is simulated in the created environment where the user is able to interact with its objects, thus enabling an ultimate immersive experience.

Augmented Reality (AR), on the other hand, is the blending of VR and real life, as developers can create images or videos within applications that are projected in the real world and blend in with real contents. Users are able to interact with the projected "virtual" contents in the real world, and to distinguish between the two.

Oculus headset, introduced in 2015, is perhaps the most publicized success story of wearable VR platforms. Other big players such as Google and Microsoft are also investing heavily in this area as VR headsets making their way to the consumer market enabling a new spectrum of applications. Needless to say, the success of VR devices is essential for the future of the wearable tech industry. This topic is discussed in details in Chapter 9.

2.7.2. Voice Recognition

Voice and speech recognition is the ability of a computer program to receive and interpret spoken words, or to understand and execute spoken commands. This technique provides a natural and intuitive way of interacting with a computer-based system allowing the user's hands to remain free.

Voice recognition technology was introduced by Bell Labs in the 1950s. However, it wasn't until the 90's of the past century when the interactive voice response (IVR) systems became a mainstream technology.

Noticeably, the integration of voice recognition in wearable technology is increasing substantially, and the market has already witnessed wearable devices that are primarily controlled by voice commands as evidenced by Apple's watch and Google Glass.

Conclusion

Advancement across the various disciplines of electrical engineering offers unique advantages and opportunities to interact with and influence our environment. This is the basis of IoT and wearable technology, and it opens up a world of new and innovative possibilities. Microprocessors, embedded sensors and other hardware elements, communication technologies and networking protocols enable advanced and well-coordinated smart wearables that improves efficiency, saves costs, and provides convenience.

References

[1] Josh Woodard, Jordan Weinstock, and Nicholas Lesher, Integrating Mobiles into Development Projects (handbook), AUGUST 2014.

[2] Analog Front End (AFE) for Sensing Temperature in Smart Grid Applications Using RTD (handbook), Texas Instruments, 2014.

[3] Terrell R. Bennett, Roozbeh Jafari, and Nicholas Gans, Motion Based Acceleration Correction for Improved Sensor Orientation Estimates, 2014 11th International Conference on Wearable and Implantable Body Sensor Networks, 2014.

[4] R. Aasin Rukshna, S. Anusha, E.Bhuvaneswarri, T.Devashena, Interfacing of Proximity Sensor with My-RIO Toolkit Using LabVIEW, - International Journal for Scientific Research & Development| Vol. 3, Issue 01, 2015.

[5] Meir Nitzan, Ayal Romem, and Robert Koppel, Pulse oximetry: fundamentals and technology update, Med Devices, vol 7, P231–239, 2014.

[6] Keiichiro Yamanaka, Mun'delanji C. Vestergaard, and Eiichi Tamiya, Printable Electrochemical Biosensors: A Focus on Screen-Printed Electrodes and Their Application, Sensors, 16(10): 1761, Oct., 2016.

[7] Khaleel, H. R. Al-Rizzo, H. M., Rucker, D. G. and Mohan, S., A compact polyimide-based UWB antenna for flexible electronics. IEEE Antennas and Wireless Propagation Letters, 11: 564-567, 2012.

[8] George Crabtree, Elizabeth Kocs, and Lynn Trahey, The energystorage frontier: Lithium-ion batteries and beyond, Materials Research Society MRS BULLETIN • Vol 40, Dec., 2015.

[9] Usama Fayyad, Gregory Piatetsky-Shapiro, and Padhraic Smyth, From Data Mining to Knowledge Discovery in Databases, American Association for Artificial Intelligence. 0738-4602-1996.

[10] Hadi Banaee, Mobyen Uddin Ahmed, and Amy Loutfi, Data Mining for Wearable Sensors in Health Monitoring Systems: A Review of

Recent Trends and Challenges, Sensors, 13(12): 17472–17500, Dec, 2013.

[11] Giuseppe Riva and Fabrizia Mantovani, Being There: Understanding the Feeling of Presence in a Synthetic Environment and Its Potential for Clinical Change, Riva and Mantovani, Intech, 2012.

[12] Ronald Azuma, Reinhold Behringer, Steven Feiner, Simon Julier, Blair MacIntyre, Recent Advances in Augmented Reality, Computers & Graphics, November 2001.

[13] George Sadowsky, James X. Dempsey, Alan Greenberg, Barbara J. Mack, Alan Schwartz, Information Technology Security Handbook, 2003.

[14] Pallàs-Areny, R.; Webster, J.G. Sensors and Signal Conditioning, 2nd ed.; John Wiley & Sons: New York, NY, USA, 2001.

[15] Huising, J.H. Smart sensor systems: Why? Where? How? In Smart Sensor Systems; Meijer, G.C.M., Ed.; Wiley: Chichester, UK, 2008; pp. 1–21. 3.

[16] Reverter, F.; Pallàs-Areny, R. Direct Sensor-to-Microcontroller Interface Circuits. Design and Characterization; Marcombo: Barcelona, Spain, 2005.

[17] Cox, D. Implementing Ohmmeter/Temperature Sensor; Microchip Technology AN512: Chandler, AZ, USA, 1994. J. Low Power Electron. Appl. 2012, 2 280

[18] Bierl, L. Precise Measurements with the MSP430; Texas Instruments: Dallas, TX, USA, 1996.

[19] Richey, R. Resistance and Capacitance Meter Using a PIC16C622; Microchip Technology AN611: Chandler, AZ, USA, 1997.

[20] Dietz, P.H.; Leigh, D.; Yerazunis, W.S. Wireless liquid level sensing for restaurant applications. In Proceedings of The 1st IEEE International Conference on Sensors, Orlando, FL, USA, 12–14 June 2002; pp. 715–719.

[21] Gaitán-Pitre, J.E.; Gasulla, M.; Pallàs-Areny, R. Analysis of a direct interface circuit for capacitive sensors. IEEE Trans. Instrum. Meas. 2009, 58, 2931–2937. 9. Reverter, F.; Gasulla, M.; Pallàs-Areny, R. Analysis of power-supply interference effects on direct sensor-to-microcontroller interfaces. IEEE Trans. Instrum. Meas. 2007, 56, 171–177.

[22] Hanes, David. IoT Fundamentals: Networking Technologies, Protocols, and Use Cases for the Internet of Things, 1st edition, Cisco Press, 2017.

[23] L. Atzori, A. Iera, G. Morabito, The internet of things: a survey. Comput. Netw. 54(15), 2787–2805, 2010.

[24] M. Mohammadi, M. Aledhari, A. Al-Fuqaha, Internet of things: a survey on enabling technologies, protocols and applications. IEEE Commun. Surveys Tuts. 17(4), 2347–2376, 2015.

[25] J. Guth, U. Breitenbucher, M. Falkenthal, F. Leymann, L. Reinfurt, Comparison of IoT platform architectures: a field study based on a reference architecture, Cloudification of the Internet of Things (CIoT), Paris, 23–25 November 2016.

[26] B. Varghese, R. Buyya, Next generation cloud computing: new trends and research directions. Future Generation Computational Systems, 79, 849–861, 2018.

[27] Q.Z.S.A.H. Ngu, M. Gutierrez, V. Metsis, S. Nepal, IoT middleware: a survey on issues and enabling technologies. IEEE Internet Things J. 4, 1–20, 2017.

[28] M. Mukherjee, I. Adhikary, S. Mondal, A.K. Mondal, M. Pundir, V. Chowdary, A vision of IoT: applications challenges and opportunities with Dehradun perspective. Adv. Intell. Syst. Comput 479(4), 553–559, 2017.

[29] https://developer.arm.com/products/processors/cortex-m

[30] M. Bilal, "A review of internet of things architecture", technologies and analysis smartphone-based attacks against 3D printers. arXiv preprint arXiv:1708.04560, 1–21, 2017.

[31] M. Botterman, For the European Commission Information Society and Media Directorate General, Networked Enterprise & RFID Unit – D4. Internet of Things: An Early Reality of the Future Internet, Report of the Internet of Things Workshop, Prague, 2009

[32] I. Toma, E. Simperl, G. Hench, A joint roadmap for semantic technologies and the internet of things. in Proceedings of the Third STI Road mapping Workshop, Crete, 2009

[33] Ahson, S.A., Ilyas, M, Near Field Communications Handbook (Internet and Communications) (CRC Press Taylor and Francis, 2011, 23 September). ISBN-10: 1420088149

[34] E. Ho, T. Jacobs, S. Meissner, S. Meyer, M. Monjas, A.S. Segura, ARM testimonials, in Enabling Things to Talk, (Springer, Berlin, Heidelberg, 2013), pp. 279–322, 2013.

[35] Haiader Raad, Fundamentals of IoT and Wearable Technology Design, Wiley/IEEE Press, 2021.

[36] F. Xia, Wireless sensor technologies and applications. Sensors 9(11), 8824–8830, 2009.

[37] B.M. Lee, J. Ouyang, Intelligent healthcare service by using collaborations IOT personal health device. Int. J. BioSci. BioTechnol. 6(1), 155–164, 2014.

[38] F.J. Kang Lee, P. Lanctot, Internet of Things: Wireless Sensor Networks (International Electrotechnical Commission, 2017.

CHAPTER FOUR

PRODUCT DEVELOPMENT AND DESIGN CONSIDERATIONS

1. Introduction

The world of wearable technology is rapidly growing and steadily pushing for new innovative products. If these devices did not provide the potential of an immense value at a low cost, there wouldn't be discussions about developing solutions based on these technologies in the first place. In fact, the demand is ongoing and the market is very exciting; however, product engineers and designers face new challenges and design constraints.

With more connected devices coming to market every day, it's extremely important to ensure their functionality, security, and interoperability. Whether creating a new smart connected product or incorporating a new technology into existing products, there are key considerations to make. For example, wearables are generally characterized by portability, flexibility, and multi-functionality compared to handheld devices. Moreover, particular performance capabilities must be integrated into compact form factors. Therefore, the design process, materials selection, and manufacturing and packaging methods could be quite unconventional at times, and need to be addressed and evaluated.

This Chapter discusses the development process and design considerations that developers must follow to guarantee a successful launch of wearable products.

2. Product Development Process

When developing a new product, there is a set of steps that must be followed to turn an innovative idea into a product available for sale. Some steps can have multiple iterations, which is typical when it comes to developing technically complex products based on hardware and software systems.

2.1 Ideation and Research

The development process starts with identifying an idea for a new product, which could be either an enhanced version of an existing product, or a non-existent product driven by a need. This step is typically followed by research and feasibility study which involves identifying the technology, materials, and methods to realize the end product.

2.2 Requirements/Specifications

The outcome of the previous step results in particular design specifications, a set of engineering requirements, along with an estimate of the cost of the end product. Section 3 of this Chapter discusses the general product requirements in details.

2.3 Engineering Analysis

Engineering analysis involves the application of scientific principles and analytical methods to understand and analyze the properties and mechanism of a system. This is enabled by breaking a system down into basic components to understand their features and relationships to each other. In the development of modern electronic devices development, this step is generally divided into three sections:

- Hardware\Electrical design
- Software\Embedded System Design
- Mechanical Design

A large number of activities take place during this phase of the project. Many of them need to be carried out and coordinated in parallel. It should also be noted that when developing an electronic device, it is important to develop a test and production strategy alongside[17]. The realization and acquiring of the equipment needed in the design and development process may take place during this stage.

2.3.1 Hardware Design

This is often the main emphasis of the development process. It begins with the top-level design, and then the requirements are broken down into smaller subsections. At this stage an electrical schematic diagram is created, the layout for a printed circuit board (PCB) is designed, and a first draft of Bill of Materials (a list of component to be used in the product development) is generated.

---

[17] ***Design for Manufacture (DFM)*** *is the aspect of the design process where consideration is given to ensure ease of manufacturing processes aiming at minimizing the production cost.*

Controls, functionalities, and user interface design are all designed in this step.

2.3.2 Software Development

Determining the operating system platform and the requirements for the device's software is a crucial step in the development process. Well defined software specifications at this step will not only reduce the number of iterative test cycles, but will also provide a clear perspective of what the essence of the product is. This step should start with the top-level design, breaking down the requirements to smaller tasks that can be tackled separately.

One of the most important decisions to make before product development is determining whether an operating system will be used. The choice of real-time operating system (RTOS), or high-level operating system (HLOS), development packages, programming languages, use of third party libraries such as networking stacks, and GUIs must be determined as well. Such decisions will have an impact on selecting the microprocessor and memory of the product.

2.3.3 Mechanical Design

The mechanical design is also an important step in the overall device development process. It does not only deal with designing a mechanical enclosure or packaging, but important aspects such as heat flow and cooling analysis, force distribution, and mechanical interfaces are all evaluated in this step. This is usually done using a mechanical modeling and simulation software packages.

It should also be noted that for those working in regulated industries, it is even more critical to use Product Lifecycle Management (PLM)[18] to push requirements that comply with regulatory and safety standards.

Next, schematic diagrams of each block of the system of the electronic design are laid out. Once schematics are ready, the design for the actual Printed Circuit Board (PCB) is created. The PCB serves as the physical platform that holds and connects all of the integrated circuits and electronic components. Moreover, code pieces are verified, and all product materials are determined in this step.

PCB Design

Generally, the PCB realization is a major step of the electronics hardware development, and is often, the most time consuming. Signal integrity analysis is also conducted as part of this step[19].

---

[18] ***Product Lifecycle Management (PLM):*** *The development of IoT and wearable devices requires electrical, mechanical, and software design teams to collaborate together beginning from the earliest stages of the project. Product lifecycle management (PLM) solutions are particularly designed to help bring all teams and designs together into a single system to enable faster design approvals, and improved traceability from concept to final product launch.*

[19] ***Signal integrity*** *deals with the electrical performance of the wires, conductive tracing and other structures used to carry signals within an electronic product. At high bit rates (high frequency clock), various effects can degrade these signals to the point where errors take place, and the system could fail. Signal integrity engineering deals with*

IoT and wearable products come in all shapes and sizes, and to meet the form-factor and ergonomic requirements of specific applications it becomes inevitable to use multiple PCBs.

Some of the drawbacks that come with using multiple PCBs are:

- Occupying additional space
- Introducing additional point(s) of failure
- Introducing additional assembly steps
- Additional costs due to PCB connectors and cables

To overcome such drawbacks, flexi-rigid PCBs are increasingly being used nowadays. These PCBs utilize a flexible polyimide layer embedded in the stack-up to hold interconnecting copper layers between the different sections allowing the finished assembly to be folded. Components are then mounted on the rigid sections in the traditional way, but the flexible sections.

Obviously, PCBs are designed using a PCB CAD software packages and hardware tools. Also, many modern professional PCB CAD systems support 3D mechanical designs to be imported for a more realistic consideration. It is worth noting that if any RF component, such as the antenna feeding element, is incorporated in the PCB then the CAD system must support RF design parameters, in particular impedance matching and return loss.

---

analyzing and mitigating such effects. It is an essential task at all levels of electronics packaging and assembly.

2.4 Prototyping

Prototyping generally refers to creating a sample model of a new product or process for testing and evaluation purposes. Prototyping provides specifications for a physical, working system rather than a virtual, theoretical one.

Some preliminary prototypes are basic and simple, and intended to visualize how a product might work, while others represent an actual representation of the end product. The selected order of a prototype, which is cost-dependent, must fit the specific requirements of the tests. The device enclosure, user interface designs, and the software deployment, which was defined in the previous step, are also implemented in this phase.

2.5 Testing and Validation

Testing and validation involve evaluating the prototype to determine if the product satisfies all the requirements and specifications defined in the second step. Testing is carried out at the very end of the hardware and software development process and is usually assigned a relatively smaller amount of time[20].

Most testing and validation procedures encompass the following:

---

[20] ***Regulatory Pre-compliance Testing***: *The goal of this testing is to detect in the early stages if there are EMC or safety issues that need to be fixed. Preparation from the development team who typically support a variety of test modes is required for both pre-compliance and official compliance testing.*

2.5.1 Review and Design Verification

At this stage, all the processes carried out previously are reviewed before starting the actual testing. This test is often called the Test Readiness Review (TRR).

Design verification ensures that the theoretical design meets the product requirements within the expected manufacturing and component tolerances. Specific areas of inspection typically include:

- Power supply check: Ensuring that voltages, currents and power dissipations are correct. Supply rail sequencing and reset timings must also be checked.

- High speed interfaces: Checking USB, Serial Advanced Technology Attachment (SATA) and memory interfaces.

- RF subsystem: Ensuring RF components are working in the design within their recommended operating conditions.

2.5.2 Unit Testing

Unit testing involves taking individual components and modules of the product, isolating it from the rest, and making sure it is functioning exactly as intended. This step is essential to ensure that each component of the main sections (software, hardware, and mechanical) meets its specifications to prevent issues during the final assembly.

2.5.3 Integration Testing

In this step, components and modules that have been tested individually are integrated and tested as a single unit. This step ensures that the interface and interaction between all parts of the device are appropriate and error-free. Lifecycle testing may also take place at this point[21].

2.5.4 Certification and Documentation

Safety and compliance certifications, such as FCC, FDA, and CE are generally required for new electronic products. These certifications provide verification that all the regulatory compliance requirements have been satisfied in order for a product to be distributed and sold legally. For example, in Canada and Europe, electronic products require both Electromagnetic Compatibility (EMC) and Safety Testing before they can be marked for sale, while this is regulated by the Federal Communications Commission (FCC) in USA. At this stage of testing, the class of the device must be defined and appropriate testing organizations must be identified.

---

[21] *Life Cycle Testing: typically involves testing the product under operating conditions significantly beyond the norm. This type of testing in the design phase is commonly known as highly accelerated life testing (HALT), and when conducted on production samples is known as highly accelerated stress screening (HASS).*

2.5.5 Production Review

A production readiness review should be conducted before the product is forwarded to production. This review marks the last step of the testing phase which is intended to ensure that the product development has been satisfactorily completed.

2.6 Production

The product can enter the production phase once a production readiness review is completed. The purpose of this stage is the industrial production of the device and making it available for purchase to the end user. Figure 2 depicts a general product development process diagram for modern electronic devices.

3. Wearable Product Requirements

When pursuing a new product development, it is essential to define the requirements which are typically captured from trial users, end customers, or market assessments.

The requirements are typically documented and serve as an agreement between the client and the product engineering team. The document can be used towards the product delivery time as a checklist for product completeness upon delivery.

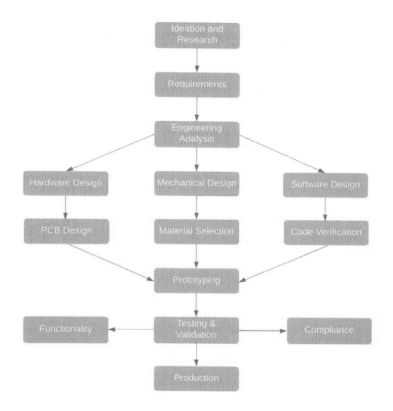

Figure 2: A general modern electronic product development process diagram

While wearable technology could help create greater user experiences and better customer satisfaction, there also exists a need to handle the requirements that define how the capabilities of the design will work in the first place. On the other hand, in the rush to get innovative products to the overly competitive market, some key requirements can be overlooked, putting security and other design features at risk. Below are some major requirements the product design/development team should pay attention to:

3.1 Form Factor

Form factor is a hardware design aspect in electronics packaging which specifies the physical dimensions, shape, weight, and other components specifications of the printed circuit board (PCB) or the device itself.

Although wearables have a small form factor in general, it is truly dependent on the type and the way they are worn (i.e.: rings and wristbands, versus glasses and jackets). This is also true for some IoT devices, think a compact smart metering device as opposed to a smart appliance or industrial equipment.

Devices with smaller form factors may offer reduced usage of material, easy handling and use, and typically low power consumption; however, they are typically associated with higher design and manufacturing costs in addition to maintenance constraints.

3.2 Power Requirements

Although some wearable devices operate autonomously such as the solar powered trackers, the majority are dependent on batteries as an energy source.

The choice of battery type and size in light-weight wearables is vital and is strongly dependent on the operational needs (expected operational duration, display utilization, computational power, etc.).

Another factor to be considered is recharge-ability. Typically, the battery is charged by plugging the device into an adapter or a powered USB port. However, the demand for wireless charging has increased recently since it offers additional convenience. For instance, users find it easier to just drop their charge-needing smartwatches on a charging base rather than plugging it into a wall adapter.

3.2.1 Energy Budget

As mentioned earlier, wearable products are typically powered by a battery. Wearable applications will be rendered useless, if battery life is unreliable and/or short. The capacity of a battery (typically in Ampere hour) is a measure of the amount of charge stored by the battery, and is determined by the mass of the chemically active material inside the battery. The capacity indicates the maximum amount of energy that can be delivered by the battery under specified conditions. A battery with a capacity of 2000 mAh (milli Ampere hours) means that the battery can deliver 2000 mA current within one hour, 1000 mA for 2 hours, or 500 mA for 4 hours, etc.

The power budget deals with the analysis of how much power a given device requires for operation. Here, this analysis is required to determine how long a wearable device will operate from a battery of a given capacity before turning off. This is determined by calculating how much time a device will spend in each of its operating modes and then summing the energies expended in each mode.

Example:

A primitive wearable device operated by a CR2032 battery with a capacity of =0.225Ah. The device consumes 1 ms operating time (on) for every 2 sec with a run current I_{run} of 8.2 mA and sleep current $I_{slp\ of}$ 1 µA. How long can the device be used before the battery has to be replaced?

Sleep time t_{slp} = 1.999 s

Run time t_{run} = 0.001 s

Sleep current I_{slp} = 1 µA

Run current I_{run} = 8.2 mA

Average Current $(I_{avg}) = \frac{Islp * tslp + Irun * trun}{tslp + trun}$ = 5.1 µA

CR2032 battery capacity C = 0.225 Ah

Average device current I_{avg} = Battery Capacity/Operating Time

5.1×10^{-6} = 0.225Ah/Operating Time

→ Operating time = 44117 hour

44117 hour = approximately 5 years[22]

---

[22] *Note: In practice the designer should pay attention to the battery self-discharge rate which is a phenomenon in batteries in which internal*

3.3 Wireless Connectivity Requirements

With most wearable products having one or more wireless interfaces, an important decision to make is whether or not the on board wireless systems should utilize original equipment manufacturer (OEM) modules or if they should be carried out a custom design. Another factor to consider is software support, i.e.: modules may be provided with a certified protocol stack (such as BLE, cellular and WiFi) that can significantly reduce the amount of overhead for software development and testing.

In wearable technology, there is a number of additional challenges that engineers face when designing antennas and wireless systems that do not exist in conventional wireless system design. For example, the degradation in the resonant frequency and return loss of wearable antennas need to be considered since they are prone to shift/deterioration due to impedance mismatch if the antenna unit is flexed or bent. Moreover, radiation patterns distortion and gain deterioration are also likely to occur. Another crucial constraint that needs to be accounted for is the close proximity of the antenna to the user's body, which implies two issues: degrading the impedance matching of the antenna due to the high water content (higher electrical conductivity) of the human tissues; and the increased amount of electromagnetic power deposition in the tissues, which gives rise to health hazards due to hyperthermia.

Designers must also ensure that no over-limit radiation of electromagnetic waves is taking place, which is characterized using a standard procedure known as Specific Absorption Rate

---

chemical reactions decrease the stored charge (capacity) of the battery even when not used in a circuit.

(SAR) test. It is also worth mentioning that other factors need to be considered in some situations where the antenna must withstand higher temperature, pressure, and humidity.

3.3.1 RF Design and Antenna Matching

Having a high RF radiation efficiency is extremely important in battery-powered IoT and wearable products. Without efficiency optimization driven by the impedance matching, the antenna and its RF circuitry can waste significant amounts of power. For wireless designs with antenna(s) mounted on the PCB, both the feeder(s) and the antenna(s) will require impedance matching to ensure maximum radiation efficiency and minimize signal reflection back to the transceiver.

The integration of the RF transceiver with other subsystems in close proximity within a small-form factor product poses a major challenge: electromagnetic interference. The negative impacts are summarized below:

- Interference due to the coupling of unwanted signals into the antenna and its feeding port, which compromises the range either as a result of reduced receiver sensitivity or lowered signal to noise ratio
- Electronic and thermal noise caused by the microcontroller/microprocessor, power supplies or other subsystems being coupled into RF system through their control interfaces.

The reader is referred to antenna design and RF circuits books for theory and design procedures.

3.3.2 Link Budget

Link budget is a commonly used metric to evaluate the range of a wireless system. All gains and losses from the transmitter to the receiver over the air-interface must be taken into consideration in order to calculate the link budget.

Link budget accounts for the attenuation of the transmitted signal due to propagation, cable and connector losses, radiation efficiency, in addition to gains from antenna topology, repeaters, and amplifiers. Effects of channel fading should also be taken into account, and can be manipulated by using techniques such as antenna diversity and multiple input multiple output (MIMO), and frequency hopping.

The basic equation for a link budget is based on Friis equation, and given as:

Received Power (dB) = Transmitted Power (dB) + Gains (dB) − Losses (dB)

First, one should start with the transmitted power at the source then add in the gain from antennas and repeaters. Next, the losses of the cables, connectors, and anything the transmitted signal passes through (channel) are subtracted.

Friis equation is used in telecommunications engineering, where the received power by the receiving antenna is calculated under idealized conditions due to a specific power transmitted by another antenna some distance away. Friis' transmission equation for free space propagation is given below:

$$P_r = P_t + G_t + G_r + 20\log\left(\frac{\lambda}{4\pi}\right) - 20\log D$$

where P_t is the transmitted power, P_r is the received power, G_t is the transmitting antenna gain, G_r is the receiving antenna gain, λ is the wavelength[23], and D is the distance between the transmitter and the receiver. For example, a link budget of 120 dB at 433 MHz gives a range of approximately 2 Km.

It should be noted that the decibel (dB) scale is widely used in electronics, signal analysis and communication systems. The dB is a logarithmic way of describing a ratio especially when the range is extremely wide. The ratio may be power, voltage, some intensity, etc.

When we convert a value V into decibel scale, we always divide by a reference value Vref, thus the quantity is dimensionless since it represents a ratio:

$$\frac{V}{Vref}$$

---

[23] ***Wavelength*** *can be obtained from the frequency of the electromagnetic wave: $C = \lambda \times F$, where C is the speed of light.*

The value in dB is given as:

$$V \text{ (in dB)} = 10\log \frac{V}{V_{ref}}$$

Power is normally measured in Watt (W) and milliWatt (mW). The corresponding dB conversions are dBW and dBm. The reader should be familiar with such conversions when working in this area.

For example: Sensitivity level (the threshold of receiving a signal) of a GSM receiver is 6.3×10^{-14} W which is equivalent to -132 dBW or -102 dBm; Bluetooth transmitted power is 10 mW which is equivalent to -20 dBW or 10 dBm; GSM mobile transmitted power is 1 W which is equivalent to 0 dBW or 30 dBm, etc.

Figure 3 expresses a link budget elaboration between a transmitter and a receiver.

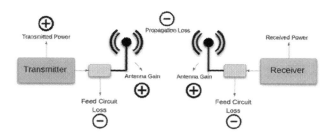

Figure 3: A link budget elaboration between a transmitter and a receiver

3.4 Cost Requirements

The cost of the product includes the initial outlay for the hardware and associated components (i.e.: sensors, microcontrollers, etc.) as well as their on-going operating costs, such as maintenance and replacement. Licensing fees for platforms, components, and device drivers should also be considered.

4. Design Considerations

Design considerations are factors that may affect the product or system requirements, design, or operational concept and should be part of the systems engineering process.

4.1 Operational Factors

Operational requirements deal with the device's essential capabilities and performance measures such as effectiveness, speed, accuracy, resolution, and consistency. Generally, the hardware and software design steps satisfy such factors.

4.2 Durability and Longevity

The device's durability depends heavily on the mechanical robustness of the packaging/enclosure material and the components quality of the internal circuitry.

The potential failure points of devices associated with structure and component deterioration should be identified and well documented. For feasibility purposes this can be achieved by testing a sample until failure (typically after a device has successfully survived the target lifetime).

For example, in some applications, smart devices have to operate in harsh environments (extreme temperatures, pressure, operation under water, vibration, corrosion, humidity, etc.). In other applications, devices could be deployed in remote, hard to reach locations making maintenance and reconfiguration very costly. Hence, rugged designs must be considered to prevent any service interruption.

It is also worth noting in wearable applications that when operating on the human body, bending, flexing and twisting become inevitable. Due to these effects, the performance of the internal components of the device could potentially deteriorate. Hence, some tests are necessary to ensure operative reliability. For example, flexibility tests are conducted by repeated trials of the prototype under bending, stretching, and twisting to monitor for any deformations, or discontinuities, and to ensure there are no wrinkles or permanent folds introduced which might compromise the functionality and aesthetics of the device.

4.3 Reliability

For a smart device to be successful it must be precise, consistent, punctual, and reliable. This is essential for the user to establish confidence and trust in the device. Any error tolerances must be identified by the manufacturer before the release of the product and must be clearly disclosed to the user. Moreover, all the system components should be accurately integrated and field tested to ensure a reliable performance. For example, the problem of Electromagnetic Interference (EMI) typically arises when multiple components are integrated in compact form factors which could negatively impact the device's performance.

EMI can also affect the accuracy of data acquisition and measurement in addition to the reliability of the communication signals. These risks can be eliminated by embedding all components within a specially-designed low-EMI enclosure.

4.4 Usability and User Interface

Usability, within the context of consumer electronics, is often defined as the ease of use, handling, and learnability of an electronic device. When designing wearable devices, it is imperative for the device to swiftly deliver the task requested by the user. Moreover, the ability to easily navigate through the device's user interface strongly influences the user's engagement and interaction.

A large number of people think of user interfaces as just software or apps on smart phones. In reality, a user interface could be anything from voice and gesture control, switches and buttons, to touch screens and control panels. Designing a highly responsive, user-friendly device ensures higher adoptability rates and pleasure of experience.

4.5 Aesthetics

Users' tastes vary based on psychological, societal, and cultural perceptions. Users also differ physically in built and complexion. To establish an emotional connection between a wide spectrum of users and the product, wearables with different styles and fashion personalities may need to be offered to the consumers.

4.6 Compatibility

As mentioned in the previous chapters, wearable devices are, in many cases, synced to a gateway (i.e.: a smart watch connected to a smartphone or tablet) for data processing and forwarding. The connection is typically carried out via WiFi, Bluetooth Low Energy (BLE) or Near Field Communication (NFC). Sometimes even when products operate under the same protocol but use different versions can cause interoperability issues. Hence, these devices must be compatible with different operating systems that could be encountered when a connection is established (i.e.: Android versus iOS).

4.7 Comfort and Ergonomic Factors

This is one of the most important design aspects in wearable devices. The device's weight, shape, size, and texture must be carefully considered.

Devices should fit users comfortably enabling usage and movement without any constraints[24]. For examples, stretch-ability, temperature and breathe-ability balance in textile-based wearables play a vital role in the commercial success of such devices.

---

[24] ***Ergonomics*** *is a relatively new discipline that deals with designing products, systems, or processes with an eye towards ensuring a proper, comfortable, and convenient handling and interaction between the device and the user.*
A proper ergonomic design requires researching other disciplines such as anthropometry, which deals with studying body sizes and shapes of a population, biomechanics, environmental physics, and applied psychology.

4.8 Safety Factors

Wearables devices are meant to enhance the quality of users' lives and must be designed with specific product safety requirements in mind. Physical and psychological harm due to misuse and/or abuse of the device, in addition to potential operation, manufacturing, and assembly hazards must be carefully evaluated. Examples include tests for device overheating, accidental electric shocks, excessive electromagnetic radiation, and material toxicity. It should be noted that the long-term physiological effects of these devices are not established yet and need to be properly researched. The psychological and social impacts of wearables will be covered in Chapter 6 of this book.

4.9 Washing Factors (Wash-ability)

Wearables that are based on fabrics are usually exposed to dirt, dust, and sweat, which might compromise their performance. Obviously, the operation and performance of wearables is required to be consistent after it is washed or soaked with water.

4.9 Maintenance Factors

The expected shelf-life of the product is often determined in the design phase. It is also determined if the product will be maintenance-free, or would require routine maintenance, and whether the maintenance is to be performed by the end user or by a professional technician.

When the product is released to the market, support provided by the releasing company is essential to evaluate and resolve any errors or bugs detected in the product. This process is also vital for improving the quality and usability of the product.

4.11 Packaging and Material Factors

This deals with the selection and assessment of the material types used in wearable devices (plastic, rubber, metal, fabric, wood, etc.). The material of the device enclosure, for example, must be strong enough to protect the internal electronic circuitry. Hence, developers must consider the mechanical factors that affect the end product quality such as: the tensile strength, density, rigidity, durability, hardness, flexibility, and stretch-ability. Further, wearable devices are typically in direct contact with the user, hence, the effects of the selected materials on the device usability such as the toxicity and biocompatibility issues must be evaluated (i.e.: skin-related allergies and irritations).

4.12 Security Factors

As mentioned previously, IoT and wearable devices collect and share data across a variety of systems and platforms. Protecting these connected devices requires an understanding of the risks and impact of cyber-attacks, awareness of vulnerabilities and the placing of plans to mitigate such scenarios without compromising design.

Moreover, functionality in many cases is a trade-off for security. Realizing a product that is secure but not practically usable can be as problematic as a product that is less secure but practical. Software, hardware, and information security must be planned carefully before a product is set up.

Related security requirements include:

- Ensuring that the product has enough memory and computational power to be able to encrypt and decrypt data at the rate they are transmitted and received

- Ensuring that the libraries of the software development support the required authorization and access control mechanisms
- Adopting standard devices that implement management protocols for securely registering new devices as they are added to a network to avoid spoofing[25], and secure security updates

4.13 Technology Obsolescence

Ensuring the availability of all components used in the bill of materials when mass production begins is of paramount importance. It is also important to estimate the availability of components during the lifetime of the product to ensure continuity. Another important consideration is the lifetime of the component itself; as a rule of thumb, it should match or exceed the expected product lifetime.

Conclusion

In a competitive industry, the development of a new product may involve risk but also creates business opportunities. The stages of product development may seem like a long process but they are introduced to save time and resources. A careful planning for a

---

[25] ***Spoofing*** *is the act of impersonating another device or user on a network in order to gain illegitimate advantage (i.e.: steal data, inject malware, or bypass access controls).*

new product development processes along with testing and validation of prototypes are essential steps towards ensuring that the new product will meet the target market needs. This Chapter discussed key engineering requirements and considerations for designing and deploying successful wearable products.

References

[1] Matthew Scarpino, Designing Circuit Boards with EAGLE: Make High-Quality PCBs at Low Cost 1st Edition, Prentice Hall, 2014.

[2] David Vogel, Medical Device Software Verification, Validation and Compliance (book), 2010.

[3] Matthew Bret Weinger, Michael E. Wiklund, Daryle Jean Gardner-Bonneau, Handbook of Human Factors in Medical Device Design, 2010.

[4] Scott Sullivan, Designing for Wearables: Effective UX for Current and Future Devices (book), 2017.

[5] Michael E. Wiklund, Jonathan Kendler, Allison Y. Strochlic, Usability Testing of Medical Devices, Second Edition 2nd Edition, 2015.

[6] Carine Lallemand, Toward a closer integration of usability in software development: a study of usability inputs in a model-driven engineering process, EICS '11 Proceedings of the 3rd ACM SIGCHI symposium on Engineering interactive computing systems, Pages 299-302, 2011.

[7] Oliver Damm, Britta Wrede, Communicating emotions: a model for natural emotions in HRI, HAI '14 Proceedings of the second international conference on Human-agent interaction, Pages 269-272, 2014.

[8] Shyamal Patel, Hyung Park, Paolo Bonato, Leighton Chan and Mary Rodgers, A review of wearable sensors and systems with application in rehabilitation, Journal of Neuro Engineering and Rehabilitation, 20129:21, 2012.

[9] Shifali Arora, Jennifer Yttri, and Wendy Nilsen, Privacy and Security in Mobile Health (mHealth) Research, Alcohol Res. 36(1): 143–151, 2014.

[10] Haider Raad Khaleel, Innovation in Wearable and Flexible Antennas, WIT Press, UK, 2014.

[11] John L. Campbell Lindsey E. Rustad John H. Porter Jeffrey R. Taylor Ethan W. Dereszynski James B. Shanley Corinna Gries Donald L. Henshaw Mary E. Martin Wade M. Sheldon, Quantity is Nothing

without Quality: Automated QA/QC for Streaming Environmental Sensor Data, BioScience, 63 (7): 574-585, 2013.

[12] Key Considerations for Successful New Product Development, Ashok. D. Morabad, HCL White Paper, May 2018.

[13] IoT hardware from prototype to production, A guide to launching hardware based IoT products for startups and scaleups, 2019, Richard Marshall, Xitex Ltd Lawrence Archard Product Development, uPBeat, Steve Hodges.

[14] https://www.newark.com/power-calculations-for-iot

[15] https://www.design1st.com/product-development-process/

[16] Balakrishnan, A. Concurrent engineering: Models and metrics. Master dissertation, McGill University, Canada, 1998.

[17] Belliveau, P., Griffin, A., & Somermeyer, S. Meltzer, R. in The PDMA toolbook for new product development, New York: John Wiley & Sons, 2002.

[18] Cooper, R. (2001). Winning at new products: Accelerating the process from idea to launch (3rd Ed.). Massachusetts: Perseus Publishing. Cooper, R., & Edgett, S. Maximizing Productivity In Product Innovation. Research Technology Management, 51(2), 47-58, 2008.

[19] Ulrich, K.T. & Eppinger, S.D. Product Design and Development. McGrawHill, 2011.

[20] Lilien, G., Morrison, P., Searls, K., Sonnack, M., & Hippel, E. Performance assessment of the lead user idea generation process for NPD. Management Science, 8(8), 1042-1059, 2002.

CHAPTER FIVE

SECURITY ISSUES AND PRIVACY CONCERNS

1. Introduction

While many of the emerging wearable technologies are giving rise to a spectrum of new applications and innovative uses, and promise super attractive user benefits, they are also posing security and privacy concerns that are largely unexplored. In fact, with the rise of big data, cloud computing, and internet of things, a new research area concerns the security of these technologies has recently emerged. Let's also not forget that wearable computing will be strongly interwoven within these technologies. Additionally, the need for wearable devices to interact and share data with a central governing communication device (for example smart watch to smart phone, medical monitoring device to a router, smart glass to smart phone, etc..) along with other sensors and peripherals would certainly create a new class of security and privacy hazards.

Typically, the interaction and data between wearable devices, gadgets, and smartphones are usually carried out through short range wireless communication protocols such as Bluetooth or ZigBee which make them susceptible to interception and potential compromise. For instance, while hackers can hijack smartphones via spyware, any connected wearables will be vulnerable too (think of intercepting a video stream from a user's smart glass which would likely contain personal and\or sensitive information).

Besides the possible risks to the wearer's privacy, such devices have the potential to be maliciously exploited by hackers to invade the privacy of others by photographing\videotaping without their consent.

Another privacy concern with the emerging smart glasses such as Google's, Lumus`, or Epson's glasses is that they allow users to simultaneously record and share images and videos of people and their activities in their range of vision in real time. This soon will trigger a greater headache when integrated with facial recognition programs which will allow users to see the person's name in the field of view, personal information, and even visit their social media accounts through the "name tag" app!

A more extreme security risk can be potentially leveraged when it comes to health applications such as hacking wearable medical equipment (i.e.: pacemakers and insulin regulators). For example, it was recently reported that a hacker could control the mechanism of an insulin regulator wirelessly from hundreds of feet away to deliver a lethal dose to a user. It was also demonstrated that a hacker could deliver a lethal voltage shock to a patient with a pacemaker. Think of other horrifying scenarios!

Compared to laptops, smartphones, and tablets, which were swiftly embraced by consumers, wearables are being adopted on a relatively slower pace. However, it's never too early to pay particular attention to the inevitable privacy and security risks that wearable devices will bring. These concerns will be more serious in the coming years as wearables gradually become a mainstream technology, and without the right privacy and security controls, data exchanged and shared by wearables could end up being used in ways never intended or even imagined.

New forms of identity theft, harassment, stalking, and fraud, are just few crimes to potentially accompany the new technology.

2. Security and Privacy Issues in Wearable Technology

The goals of information security are defined best by the CIA triad. Not to be confused with the government agency, CIA stands for Confidentiality, Integrity, and Availability. The goals of Information Security, and its largest branch, Cyber Security, are to protect the confidentiality of information, the integrity of information from unauthorized changes, and ensuring the availability of Information and systems to the users at the expected performance level. The term information and data in the CIA triad has a broad definition, and it spans from high level user information to metadata clues. A sexual orientation, for example, can be revealed by correlation of geolocation information and its recorded time all the way to a low-level setting or code that could reveal an exploitable vulnerability. Like everything else, makers of wearable technology are forced to develop their products within the boundaries of the information security goals (CIA Triad). This applies to both data in motion (during transmission), or data at rest (stored data, or system configuration).

Unlike laptops, tablet pads, and cell phones, wearable devices are often worn or incorporated into the user's body, thus providing sensory and scanning qualities that promote efficiency, productivity, and engagement, and facilitate data collection and tracking. As mentioned earlier, the processing outcome of such information provide considerable benefits such as helping the users manage their fitness routines and biofeedback, enhancing personal productivity, and making the use of other technologies more convenient. For example, a Fitbit can collect information about the number of calories burned, the Apple watch can greatly enhance the user interaction with other technologies, while a respiratory tracker can gather your breathing and coughing patterns and has the ability to predict asthma attacks.

Satisfying the demands of today's information-oriented consumers, wearable gadgets deploy various sensors to collect a wide spectrum of biological, environmental, behavioral and social information from and for their users; and obviously, the more sensors and data collectors incorporated into our bodies, apparels, and accessories, the greater the amount of sensitive information that will be transported, stored, and processed on these devices. While wearable electronics can provide the modern user countless benefits in the areas of personal communication, wellness, health care and entertainment, they also upraise security and privacy concerns. This is also confirmed by the Federal Trade Commission (FTC), which states that while wearables promote wellness, and have the ability of providing convenience and better health services, they are also "collecting, transmitting, storing, and often sharing vast amounts of consumer data," thus creating a number of privacy risks. In particular, wearable devices challenge traditional privacy principles and pose a distinct challenge to the collection, use, and storage of health, location, financial, and other sensitive information.

In a recent study, 57% of the participants surveyed believed that they would rely on wearable technology for support rather than their family and friends, 73% of individuals anticipated that wearables would make media and entertainment more enticing and fun, while 70% reported they would use employer-provided wearables to stream health-related information in exchange for discounts on their insurance premiums.

Before proposing or designing effective solutions for the privacy and security imposed by the use of wearables, we first need to identify and understand *what* are they might possibly be, and *how* they are discerned.

2.1 Privacy and Security Concerns in Digital Technologies

The topic of privacy and security concerns in technology is not new and it is certainly not limited to the digital technologies. In fact, it has been discussed as early as the 19th century by Louis Brandeis and Samuel Warren who defined in their work "The Right to Privacy" the protection of the private domain as the founding basis of individual freedom in the modern age in response to the capacity elevation of government, press, and their related institutions to invade facets of personal activities that became accessible under new technological change.

However, privacy issues related to mobile and portable technologies are relatively new, and complex to study and analyze. Furthermore, wearables that continuously collect information utilizing their peripherals and sensors such as: microphones, cameras, Global Positioning Systems, and biomedical data collectors, add newer challenges to the user's privacy.

Most of the previous research on user privacy has emphasized on mobile devices and their applications, social networks, in addition to other security concerns such as browser, malicious code, and physical access-based attacks. On the other hand, only a little is known about the privacy concerns of wearables, especially from a user-centered perspective.

For example, previous studies show that most users are not aware about *what* data are collected and *how* they are processed. For example, whenever an Apple phone asks Siri (a built-in technology that enables iPhone and iPad users of speaking voice commands in order to operate the mobile device or access internet information) a question, their voice is sent as a file to the data servers at Apple's headquarter for analysis. The file is then given a number that associates your phone to the question you asked. This data is then stored on Apple servers for up to two years for testing purposes, though your file number is deleted after six months.

On the other hand, there are mainly three points of vulnerability associated with consumer information in wearables: local storage, software-cloud communication through WiFi and cellular networks, and cloud storage. Obviously, the first area of vulnerability, i.e.: stealing information off the local storage of the device is not of high probability unless the device is lost or stolen and gets hacked into. While the information stored on the local drive of the device is more vulnerable when transmitted via Bluetooth as it could be compromised by the use of Bluetooth packet sniffers. Packet sniffing is a utility that has been in use since the early days of computer networks. Packet sniffing allows network professionals to capture data legitimately as it is transmitted over a network and is used mainly for the diagnosis of network issues. However, this utility can also enable malicious users to illegally gain access to a system or network, and to capture unencrypted data when in transit, like passwords and usernames. The data at this point is not readable until it has been gone through un-packaging processes and decoded by the corresponding software used by the transmitting device. Needless to say, the type of the available information will predominantly determine the functionality within the wearable device. For example, a smart watch with a social media notifications feature may require login data to be stored and periodically checked on the smart watch.

That feature may expose the device to a higher risk compared to a simple fitness tracker.

The second point of vulnerability is when the software communicates with the cloud through a WiFi or cellular connectivity. As reported by Symantec: the data would be susceptible to a number of possible threats when transmitted which includes traffic sniffing; man-in-the-middle (MitM) attack, where the attacker stealthily relays and possibly modifies the communication between two users who are believed to be communicating directly with each other; and data redirection attacks, which could forward the data to a malicious server.

When the software communicates with the cloud, there is often more data to be compromised since the data transfer from the local drive of the device to the application is typically accompanied by personal information such as: name, e-mail address, phone number, and even location to guarantee that the data is being forwarded to the intended account. The majority of wearable industries are aware of the severity of this issue if such sensitive data is compromised, and their effort is focused on minimizing these risks by applying robust encryption and authentication methods on the data under consideration.

2.2. Threats and Attacks

As The sophistication of cyberattacks targeting IoT and wearable devices is on the rise. In fact, 2016 witnessed the emergence of an IoT-based botnet[26] that almost paralyzed the internet.

---

[26] *A **botnet** is a number of devices connected to the Internet that can be used to perform Distributed Denial-of-Service (DDoS) attacks, access device data, send spam, etc.*

Besides botnets, the following years witnessed a growing increase in malwares, and cryptominers that can be used to target cryptocurrency. It is also worth mentioning that cloud storage poses threats with high impact due to the large amount of Personally Identifiable Information (PII). Moreover, attacks on the cloud are normally sophisticated, coordinated, planned well ahead of time and are conducted by professional cyber attackers.

The motivations behind cybercrimes can be quite simple: money and information. According to one study, financial and espionage-driven motivations make up about 93% of attacks. Attacks driven by large-scale disruption and destruction such as terrorism and revenge attacks are amongst other motivations.

Some of the most notable attacks including malware based (such as Mirai, Satori, and VPNFilter), exploit kits, advanced persistent worms such as Stuxnet, in addition to cybercriminals and state-sponsored activities which are exploited to access sensitive data, instill operational, or cause a reputational damage are possible due to system vulnerabilities such as:

a- Weak authentication mechanisms
b- Weak password standards
c- Poor cryptography implementation which promotes man-in-the middle attacks, and session and protocol hijacking

2.3. Threat Modeling

To better illustrate the threat landscape in the realm of IoT and wearable technology, we can use Threat Modeling.

Threat modeling is a structural and systematic way by which potential threats can be identified, prioritized, and eventually mitigated. The objective of such modeling is to provide cybersecurity teams with a systematic perspective of the potential attacker's profile, attack vectors, and the assets most likely to be targeted by an attacker.

There are various ways and methodologies of establishing threat models, we will adopt one model created by Microsoft, called *STRIDE*:

Spoofing: A spoofing attack takes place when an attacker pretends to be someone they're not. An attacker may be able to pull out cryptographic key data from a device, then accesses the system with another device using the stolen identity of the device the key has been taken from.

Tampering: Tampering refers to maliciously modifying processes, data at rest, or data in-transit. An attacker may partially or fully replace the software running on the device, which can potentially have adverse consequences if the attacker is able to add or remove some functional elements, or modify or destroy important data.

Repudiation: Repudiation refers to denying that an activity or an event has taken place. Attackers often try to hide their malicious actions, to avoid getting detected. For example, they might try to erase their illegal activities from the logs, or spoof the credentials of another user.

Information Disclosure: Information Disclosure refers to data leaks or data breaches. Attackers may try to run a modified software on the compromised system which could potentially help leak data to unauthorized parties. In IoT and wearable applications, the attacker may try to gain access into the communication path between the device and gateway, or gateway to cloud to extract information.

Denial of Service: Denial of Service refers to degrading or denying a service or a network resource to users. In some cases, attackers would benefit from preventing users to access a system, for example as a way to blackmail and coercion.

Elevation of Privilege: Elevation of Privileges refers to gaining access to resources that one is not allowed. Once a user is identified on a system, they typically have some form of privileges, for instance, they are authorized to perform some actions or access some resources, but not necessarily all of them. Therefore, an attacker might attempt to gain additional privileges, for example by spoofing a user with higher privileges or by tampering the system to upgrade their own privileges.

2.4. Common Attacks

The most vulnerable area when it comes to data theft in wearables lies within cloud storage. As reported by Symantec, there could be several risks involved based on the configuration of the system which include:

1- SQL injection attacks: an attacker can use an SQL (Structure Query Language) injection to bypass the authentication and authorization mechanisms of a web application mainly to access the contents of a database. It can also be used to modify the records in a given database which affects data integrity, and to provide an attacker with unauthorized access to sensitive information including: personally identifiable information (PII), which is information that can be used to identify, contact, or locate a user, customer data, intellectual property and other sensitive information.

2- Account brute force login attacks: is one of the oldest and most common attacks performed against Web applications. The goal of a brute force attack is to penetrate user accounts by repeatedly attempting to guess the password of a user.

3- Distributed denial-of-service (DDoS) attacks: a DDoS attack is an attempt to force an online service to be unavailable via overwhelming it with excessive traffic. Targets can range from banks and retailers to news websites.

4- Back door attacks: A backdoor attack is a way of accessing a computer program through bypassing its security mechanisms. A programmer may intentionally install a back door so that the program can be accessed for troubleshooting purposes. However, hackers often make use of back doors as part of a malicious exploit.

5- Default Password Attacks: are the most common large-scale attacks. Failing to change the vendor's default Password is a substantial security risk. It is worth noting that some computing devices such as routers typically come with a unique default password printed on a sticker (affixed on the back of the device), which is considered a more secure option. However, some manufacturers derive the password from the device's MAC address using a standard algorithm, in which case passwords could be easily reproduced by cyber criminals.

AS mentioned above, cloud storage is considered the most vulnerable threat due to the large amount of Personally Identifiable Information. Virtually any user with internet connectivity could gain access to the cloud and steal a company's data collection. However, attacks on the cloud are normally sophisticated, coordinated, planned well ahead of time and are conducted by professional cyber hijackers.

As argued by Swat Solutions: "Does the wearable world pose a greater risk to data compromise than other technologies?" They report: "The data collected by wearables isn't any more comprehensive than the information collected by financial institutions, healthcare systems, Facebook, Twitter, or LinkedIn. As of March, 2015, no publically acknowledged data breach has occurred in the wearable community but it may be simply a matter of time." Their prophecy came true in January 2016 when multiple online accounts belonging to Fitbit users have been compromised by hackers who were able to change the email addresses and usernames and also tried to defraud Fitbit out of replacement products under a user's warranty. The hackers were also able to gain access to the Fitbit users' GPS history, which shows where and when the user regularly runs\cycles, in addition to information about the user's sleeping time and pattern.

A few examples of the data breaches reported in 2014 and 2015 include: Anthem Health, JPMorgan Chase, Sony, Target, Apple iCloud, Home Depot, and Goodwill. In the cases of JPMorgan Chase, Anthem Health, and Target, hackers were able to steal extremely sensitive information such as: social security and credit card numbers, financial data, and health information. These companies utilized state of the art security protocols to protect their data but were still broken into. Such security concerns should steer the attention of wearable designers and manufacturers to the lessons learned by the big companies' failures outside the wearable world. As with all technologies that involve data collection, transmission, and storage, users should be extra cautious with their wearable devices especially when handling sensitive data such as in online banking, government-related and personal applications, or even when checking social media accounts on a public network.

2.5. Privacy Issues

In one study, Symantec found that all wearable fitness trackers under study were vulnerable to location tracking. Location information can raise serious privacy concerns since a record of a user's movements is disclosed. While this data can be useful for personal use, users are not always aware of where this data is forwarded to and with whom it may be shared. For example, the same study found that 52% of fitness tracking apps do not incorporate privacy policies or statements concerning data collection and use.

In 2014, the FTC Bureau of Consumer Protection addressed the potential privacy risks of geolocation data collected form wearables and discussed how location data could potentially expose highly personal information about an individual, such as whether a user visited an AIDS clinic, a hospital, or a worship facility. Clearly, such data can be misused if accessed by malicious attackers, traded with companies, or gathered by stalking apps. Geolocation data can also facilitate criminal activities such as stalking, robbery, and even kidnapping, as such data can easily pinpoint a user's current or future location.

Another study conducted by FTC on twelve health-related apps, showed that sensitive health conditions such as pregnancy status, gender, and ovulation information was transmitted to 76 third parties including advertisement and analytics firms. While a recent study by EMC Corporation estimates that the information obtained from health records hold about fifty times more value than credit card information as such information can be easily used in identity theft and other fraudulent activities.

2.6. Potential Solutions

The data security solutions in wearable technology are mainly concerned with authentication and encryption issues. As mentioned previously, the most common security vulnerabilities found in wearables come from insecure wireless transmission of data (Bluetooth and WiFi) from the wearable's local storage, to the cloud or smartphone, and vice versa.

Studies show that the majority of new technology users disregard the risks of malicious attacks. A common sense solution here is to consider a strong and diverse passcodes for devices that are password-enabled rather than using the default one set by the vendor. This could greatly reduce the amount of security breaches from easy-to-hack passcodes. Users should also consider updating their wearables and smartphone apps regularly to ensure that the newest security updates are in effect.

Establishing an encrypted link in sensors would also decrease eavesdropping and proxy attacks in wireless communication schemes, especially in BLE. Moreover, developers of wearable devices should consider implementing a built-in security mechanism and protection features such as user authentication or PIN system in addition to an encrypted data storage capability.

Biometric user authentication, on the other hand, is a possible solution to the abovementioned threats. Application of biometrics in the computing and communications area is not new, hence it would be a natural candidate for implementation in wearable technologies. Moreover, biometric authentication offers more convenience in wearable devices, especially in compact platforms (i.e.: Passwords and PIN codes would be less appropriate). Also, from the Internet of Things' perspective, wearable biometrics is regarded by many researchers to be a universal authentication platform when users interact with interconnected devices. What is meant by a universal authentication platform, is that a network of devices that 'instinctively' knows who you are, where you are, and what you need. However, this solution will cover specific aspects of the raised privacy and security concerns, and will certainly triggers some more.

When it comes to users' privacy, the European Parliament (EP) recommends passing the privacy Impact Assessments based on the RFID applications privacy and data protection framework to wearables. The EP also recommends deleting raw data once processed, and immediately informing the user once a data compromise risk is detected.

In a workplace environment, storing and transferring sensitive data using wearables could violate privacy laws such as The Health Insurance Portability and Accountability Act (HIPAA), or the firm's Intellectual Property. One proposed solution from a business leader is to create new organizational rules and update the firm's network security infrastructure in order to detect and potentially control the data traffic from and to wearables. Another possibility is to consider implementing a Mobile Device Management (MDM) system to manage what features are enabled or disabled on the wearable device and the smartphone. Furthermore, advanced security solutions could enable the analysis of data flows and identifying the type of device transmitting any data. For examples, in platforms where wearable devices exist, a network administrator could be alerted when an out of network data communication takes place. Even if this technique may not be able to block the communication, detecting the transmission generated from the wearable device may be sufficient to inform a network administrator that an unauthorized device is being used on the network. [27]

---

[27] ***The Health Insurance Portability and Accountability Act (HIPAA)*** *generally sets the US standards for protecting health information, which may consist of electronic and other forms of media containing identifiable information concerning an individual's past, current, or future physical or mental health that is generated or received by a healthcare providers or employers.*

References

[1] Internet of Things, Privacy & Security in a Connected World, FTC Staff Report, January 2015.

[2] Samuel D. Warren and Louis D. Brandeis, The Right to Privacy, Harvard Law Review, Vol. 4, No. 5, pp. 193-220, Dec. 15, 1890.

[3] Ke Wan Ching and Manmeet Mahinderjit Singh, Wearable Technology Devices Security nad Privacy Vulnerability Analysis, International Journal of Network Security & Its Applications (IJNSA) Vol.8, No.3, May 2016.

[4] Neil Bergman, Jason Rouse, Hacking Exposed Mobile: Security Secrets & Solutions 1st Edition, McGraw-Hill, 2013.

[5] Robert P. Hartwig, Claire Wilkinson, Cyber Risk: Threat and opportunity, October 2015.

[6] Janice Phaik Lin Goh, Privacy, Security, and Wearable Technology, Landslide, ABA Section of Intellectual Property Law, Volume 8, Number 2, 2015.

[7] Kahina Chelli, Security Issues in Wireless Sensor Networks: Attacks and Countermeasures, Proceedings of the World Congress on Engineering 2015 Vol I, WCE 2015, July 1 - 3, 2015.

[8] Biometrics: Enhancing Security or Invading Privacy?, The Irish Council for Bioethics, 2009.

[9] Ovidiu Vermesan, Peter Friess, Internet of Things: Converging Technologies for Smart Environments and Integrated Ecosystems, 2013.

[10] Privacy and Data Protection Impact Assessment Framework for RFID Applications, 12 January 2011.

[11] Kazuo Yano, Koji Ara, Junichiro Watanabe, Satomi Tsuji, Nobuo Sato, Measuring Happiness Using Wearable Technology, Technology for Boosting Productivity in Knowledge Work and Service Businesses, Hitachi Review Vol. 64, No. 8, 2015.

[12] Internet of Things, Privacy & Security in a Connected World, FTC Staff Report, January 2015.

CHAPTER SIX

PSYCHOLOGICAL AND SOCIAL IMPACTS

1. Introduction

The 21st century is witnessing substantial technological changes which have given rise to new social and psychological implications.

Along with the growth that wearable technology has undergone in the past few years come new behavioral trends which may be indicative of social and psychological influences amongst users. Hence, it is never early to address and consider the potential adverse impacts of such effects and influences created by both current and foreseeable trends.

For example, while it is widely accepted that smartwatches greatly enhance our connectivity, and to attain regular tasks more conveniently, we have to bear in mind that such technology is becoming increasingly invasive. It is also important to consider the degree to which this could elevate stress levels and anxiety.

The average smartphone user, according to, unlocks their device 110 times a day, a number that would undoubtedly increase once wearables become more mainstream. Another study shows that the current generation spends about half of their time thinking about something other than what they are intended to be doing which affects their happiness and satisfaction levels. This is largely due to the device's distraction which diverts an individual from being immersed i a genuine experience, such as being sincerely engaged in a conversation, appreciating a scenic view, or enjoying a good food, to focusing on the continuous interruptions from a phone's text, app, or a social media notification.

Despite the many lifestyle benefits associated with wearable technology, we need to also consider its potentials to disrupt users' lives and the adverse effects on their social and mental wellbeing, which will be the discussion topic of this chapter.

2. The Psychological Effects of Wearables

While it is generally agreed that modern communication technologies are useful and helpful, and make the lives of people easier, they may undeniably make us restless, anxious, subject to frequent distraction, and always in need for constant entertainment. What is not helping is that the pace at which technology is progressing is so fast that our psychological processes are not keeping up.

In a book titled "iDisorder" by Larry Rosen, the author hypothesizes that many technology users today could be diagnosed with what he calls an iDisorder. Rosen describes the psychological disorder as follows: "An iDisorder is where you exhibit signs and symptoms of a psychiatric disorder such as OCD (Obsessive–Compulsive Disorder), narcissism, addiction or even ADHD (Attention Deficit/Hyperactivity Disorder), which are manifested through your use, or overuse, of technology". A compulsive desire to check for text messages or emails, a desperate need to constantly update your Facebook status, or an obsessive addiction to iPhone games are all indications of iDisorder. There is no doubt that technology is affecting the way our brains function, whether or not these behavioral changes due to technology use are classified as a disorder, or a form of mental illness.

Smartphones have brought a groundbreaking level of convenience to people's lives. On the flipside, they make us accessible at any minute to colleagues, employers, friends, and relatives which may not always be ideal.

The way in which technology now is integrated within ourselves means that it is becoming harder to achieve a work-life balance. The key challenge today is about attaining the willpower to withstand checking or responding to work-related emails while at home or on vacation. With more enterprises implementing wearable technology, new workplace culture would further lengthen and intensify the working day, which may turn the minor mental issues into real mental health problems.

For example, in the case of Swisscom Chief Executive Carsten Schloter's suicide, media reports suggesting that he had become dangerously addicted to his smartphone. While this may represent an extreme outcome of the pressures that being constantly connected can bring, psychological studies confirm that smartphones are indeed introducing a new form of stress for users at home, at work, and in social environments.

To prevent a new age of connected workplaces from further affecting the employee's mental health, employers will need to invest in tracking the extent of technology-related mental health issues, such as email addiction, that arise across the workplace. Employees need to be educated about the potential risks associated with using wearable technology, including the potential of being addicted to the device. This can be done through seminars and workshops offered at the workplace as a preventative measure as well as providing resources for employees that may have developed an addiction already.

A new report from the family technology education group, Common Sense Media, shows that more than a third of children under the age of 2 use some form of mobile media. While another report confirms that as children age, this percentage dramatically increases with 95% of U.S teens (12-17) spending around nine hours online on average, while the average is about six hours per day for kids between the ages of 8 and 12.

A study published in Psychology Today reports that the use of technology can modify the actual wiring of the brain. The time spent with technology alters the way teens' brains work. For instance, the study shows that while videogames may condition the brain to pay attention to several stimuli simultaneously, they may lead to distraction and reduced memory. Kids who constantly use search engines may excel at finding information, but not good at remembering it. Moreover, the study reports that kids who use technology excessively may not have sufficient opportunities to use and develop their imaginative side or to read and deeply reflect about a given material.

Results of a recently published research shows that higher levels of texting is correlated with poorer quality of sleep more likely because study subjects felt obliged to respond to texts received during the night. Furthermore, excessive texting activity was correlated with elevated difficulties in stress management for those already experiencing some form of stress.

It can certainly be predicted that the use of wearable devices that enable hands-free operation and immediate access to unlimited information will likely promote the severity of the abovementioned issues.

3. Social Implications

Although the majority of people enjoy a moderate use of social networking, a percentage of users have problems controlling the amount of time spent online. Many psychologists relate the urge to visit social media sites to addictive behaviors. Users who are addicted to social media often get to prefer online communication over face to face communication. They spend a disproportionate amount of time on their smartphones, tablets, or computers because it enables them to control social interactions and avoid many of the possible uncertainties involved in direct face to face contact.

The social media addiction or overuse has intensified with the widespread availability of internet-enabled mobile phones and other handheld devices. Many users visit their social media accounts while walking down the street, attending a business meeting, or dining at a restaurant, taking advantage of the unconstrained connectivity. The emergence of wearables gives users even more accessibility to social media sites which would undoubtedly further intensify the addictive side effects. In fact, social researchers and professionals warn that some users may become very dependent on online social media that they neglect essential aspects of their off-line existence, such as their jobs, family, close friends, and health.

A recent study conducted on two groups of sixth graders revealed that the group who abstained from the use of any electronic devices for five days achieved better scores at picking up on emotions and nonverbal gestures of photos of faces than the other group that used electronic devices. The increase in face to face interaction that the first group had made them more sensitive to facial expressions. The study concludes that the use of technology can affect a child's ability to empathize.

Another study found that kids who use videogames and online media for more than four hours a day do not have the same perception of wellbeing compared to those who used that technology for less than one hour. Many social experts agree that with less physical contact, kids will have difficulty developing social skills and emotional responses.

The increasing use of mobile devices and their wearable complements (i.e.: smart watch) at events, restaurants, and other social venues has generated a substantial criticism as well. Hence, it is reasonable to deduce that the use of such devices to access social media in many cases would serve to isolate people, rather than connecting them.

4. Technology Acceptance Factors

It is largely accepted that understanding the importance of social and psychological influences of the targeted users before designing a product itself determines its adoptability.

For example, supportive wearables can immensely increase individuals' independence; however, if the user's needs and expectations are not met due to a minor technical issue it might lead to a higher rate of abandonment.

In the case of elderly alarm systems that are based on necklaces and wristbands, it was found that some individuals are ashamed of wearing it and prefer not to use these systems to avoid looking old or dependent. Aesthetics and appearance, on the other hand, poses a great influence on the adoption of wearables. Self-image is based on how people view us and may affect how we feel about ourselves, which in turn affects self-confidence. For example, a study found that a trendy heartrate monitor wristband may be accepted whereas an alarm device that looks a lot like a medical aid may be deemed stigmatizing by an elderly user.

When it comes to personal privacy, users tend to be sensitive toward publically sharing information about themselves especially if it brings potential harm, socially. Privacy is an extremely critical issue, especially in a mobile and wearable computing environment. For example, a smart jacket that displays the user's body conditions and emotions via special graphics has been strongly rejected. Surveyed potential users exhibited unwillingness to disclose their emotions publically because they considered them to be highly personal information and potentially harmful to expose.

In conclusion, it is clear that many wearables will revolutionize the 21st century, and will continue to permeate our society. At the same time, it is very probable that, like earlier technological trends, these gadgets will introduce unprecedented social and mental health issues. Hence, it would be smart to take a proactive approach to address the psychological and social impacts of wearable technology. One way to do this is for researchers in the technical, social, and psychological sciences fields to collaborate in conducting research on the possible effects and impacts of this transformative technology.

References

[1] Kazuo Yano, Koji Ara, Junichiro Watanabe, Satomi Tsuji, Nobuo Sato, Measuring Happiness Using Wearable Technology, Technology for Boosting Productivity in Knowledge Work and Service Businesses, Hitachi Review Vol. 64, No. 8, 2015.

[2] Dennis L. Murphy, Kiara R. Timpano, Michael G. Wheaton, Benjamin D. Greenberg, Euripedes C. Miguel, Obsessive-compulsive disorder and its related disorders: a reappraisal of obsessive-compulsive spectrum concepts, Dialogues Clin Neurosci, 12(2): 131–148, 2010.

[3] Zero to Eight: A Common Sense Media Research Study Children's Media Use in America, FALL 2011.

[4] Amanda Lanhart, Kristen Purcell, Aaron Smith, Social Media and Mobile Internet use among Teens and Young Adults,Pew Research Center, 2010.

[5] Clayton, R. B., Leshner, G., & Almond, A. (2015). The Extended iSelf: The Impact of iPhone Separation on Cognition, Emotion, and Physiology. Journal of Computer-Mediated Communication, n/a-n/a. doi: 10.1111/jcc4.12109.

[6] Daria J. Kuss, and Mark D. Griffiths, Online Social Networking and Addiction—A Review of the Psychological Literature, Int J Environ Res Public Health, 8(9): 3528–3552, 2011.

[7] Yalda T. Uhls, Minas Michikyanb, Jordan Morrisc, Debra Garciad, b, Gary W. Small, Five days at outdoor education camp without screens improves preteen skills with nonverbal emotion cues, Computers in Human Behavior, Volume 39, Pages 387–392, October 2014.

[8] Kaveri Subrahmanyam, Robert E. Kraut, Patricia M. Greenfield, Elisheva F. Gros, The Impact of Home, Computer Use on Children's Activities and Development, Children and Computer Technology, Vol. 10, No. 2 – Fall/Winter 2000.

[9] Cherrylyn Buenaflor and Hee-Cheol Kim, Six Human Factors to Acceptability of Wearable Computers, International Journal of Multimedia and Ubiquitous Engineering, Vol. 8, No. 3, May, 2013.

[10] Patrick C. Shih, Kyungsik Han, Erika Shehan Poole, Mary Beth Rosson, Use and Adoption Challenges of Wearable Activity Trackers, iConference, p. 1-12, 2015.

[11] Knight, J. F. and Baber, C. (2005). A Tool to Assess the Comfort of Wearable Computers - Human Factors, The Journal of the Human Factors and Ergonomics Society, 2005, pp 47- 77.

[12] Lennart Hardell, Michael Carlberg, Fredrik Söderqvist, and Kjell Hansson Mild, Long-term use of cellular phones and brain tumours: increased risk associated with use for more than 10 years, Occup Environ Med. 2007 Sep; 64(9): 626–632.

CHAPTER SEVEN

HEALTH AND SAFTY CONCERNS

1. Introduction

With the debut of products like Apple Watch, HoloLens, Google Glass, and Fitbit, more reports have surfaced discussing the potential hazards of being around continuous RF electromagnetic radiation. This is the same type of energy radiated by cell phones, tablets, and laptops. Such studies have been around for some time which allow for safety comparisons between old and new technologies.

In order to minimize health risks, most wearable products utilize Bluetooth Low Energy (BLE) technology, which emits lower levels of RF energy compared to cell phones and other WiFi enabled devices. While it is confirmed that exposure to electromagnetic radiation from wireless devices induces heating in the area where they are held, industry spokesmen argue that wearables are much safer since they radiate lower energy levels than cellular and WiFi based devices.

However, new research is beginning to reveal inconvenient truths about the non-thermal effects caused by radiation, and hundreds of experts are calling on governments to issue precautionary statements and adopt regulations that protect consumers from these risks. France, for instance, has responded to such calls, and is eliminating WiFi in day cares and pre-schools, while the city of San Francisco has introduced safety fact sheets when purchasing a new cell phone.

Despite the risks, many wearable medical devices provide unprecedented health benefits and have life-saving features as discussed in the previous chapters. At the same time there is evidence of an increased number of users developing electromagnetic hypersensitivity (which will be explained in more detailes later in the chapter) due to cumulative effects of RF radiation. Thus, the challenge for scientists henceforward seems to be in raising awareness of the potential risks the technology poses to the user's health, while balancing benefits against drawbacks.

This chapter reports the major health concerns of using wearable technology raised by research findings and health organizations, in addition to recommendations on ways to minimize such potential risks.

2. Electromagnetic Radiation and Specific Absorption Rate

Electromagnetic waves in the radiofrequency range (3 KHz-300 GHz) are what enable transmission of wireless telecommunications including cellular networks, televisions and radio broadcasting, in addition to all wearable devices. These waves are radiated by antennas which are designed with different sizes and shapes to allow different frequencies of operation and radiation patterns (which specify the direction and distribution of the radiated electromagnetic energy).

When wearables are in use, the human body absorbs some of the electromagnetic energy radiated from wireless devices. The amount of the absorbed energy is calculated using a measure called the Specific Absorption Rate (SAR), which is expressed in Watts per Kilogram of body weight.

The Federal Communications Commission (FCC) mandates that every wireless device sold in the USA must be tested and verified to have a SAR less than 1.6 Watts/Kilogram (W/Kg) before it can go on sale. Canada's regulations also mandate a 1.6 W/Kg limit, while the European Union and Australia require a more stringent 2.0 W/Kg limit.

It is worth noting that SAR measurements have received a great deal of criticism in the past few years, and many organizations consider SAR as an unreliable measure of whether or not a wireless device is safe. A mobile phone, for example, may have a SAR of 0.9 W/Kg, but that may not be any safer than a device with 1.2 W/Kg. A phone's SAR value, for example, can vary widely during a call as you alternate between transmission channels and as the distance from a cellular tower is increased.

The FCC Guide "Specific Absorption Rate (SAR) For Cell Phones: What It Means for You" states: "ALL cell phones must meet the FCC's RF exposure standard, which is set at a level well below that at which laboratory testing indicates, and medical and biological experts generally agree, adverse health effects could occur. For users who are concerned with the adequacy of this standard or who otherwise wish to further reduce their exposure, the most effective means to reduce exposure are to hold the cell phone away from the head or body and to use a speakerphone or hands-free accessory. These measures will generally have much more impact on RF energy absorption than the small difference in SAR between individual cell phones, which, in any event, is an unreliable comparison of RF exposure to consumers, given the variables of individual use". Moreover, SAR limits set by the FCC do not take into consideration that the human body is also sensitive to the power amplitudes and frequencies responsible for the microwave hearing effect, also known as the Frey effect that occurs with exposures of 400 µW/cm², which is well below the FCC SAR limits.

The Frey effect, which was first reported by individuals working in the vicinity of radar transmitters during World War II, consists of audible clicking and buzzing induced by pulsed radio frequencies. These audible clicks are generated inside the head without the need of any receiving antenna. The cause is thought to be thermo-elastic expansion of parts of the middle and inner ear.

It should be noted that many wearables like Apple Watch and Fitbit use Bluetooth Low Energy (BLE) technology, which emits much lower power than classic Bluetooth, and significantly lower than cell phones. In fact, in some cases ultra-low power devices are not required by the FCC to be tested for SAR compared to cell phones and laptops which must pass a rigorous testing. However, some types of wearables do not limit their wireless technology to Bluetooth. For example, Google Glass, Recon, and Oculus Rift use WiFi too, which is comparable to cell phones in terms of electromagnetic energy radiation.

As mentioned previously, antennas are what enable wireless communication in electronics. In wearables, antennas are required to be compact, lightweight, and mechanically robust. They are also preferred to be flexible with a low profile (thin), yet, they must express high efficiency and desirable radiation characteristics.

To minimize SAR in wearable devices, antennas are preferred to have a uni-directional (hemi-spherical) radiation pattern, radiating away from the user's body to reduce the user's exposure to electromagnetic radiation. However, antennas that offer such characteristics, like microstrips, suffer from a relatively narrow bandwidth which is a function of the platform thickness. Thus, the majority of handheld electronics designers choose printed monopole\dipole antennas which offer a simple, thin, compact, and cost-effective solution, but also exhibit an omni-directional radiation pattern (i.e.: radiates in all directions including the user's body).

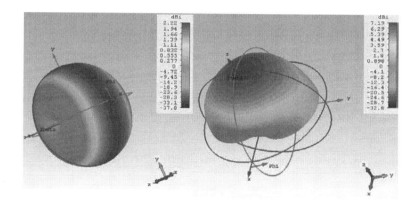

Figure 1: Omni directional radiation pattern (left), and Semi-directional (hemi-spherical radiation pattern (right)

Many antenna designs have been proposed in the literature to solve the SAR and thickness tradeoff. One design by the author of this book features a low profile printed monopole antenna integrated with a compact Artificial Magnetic Conductor (AMC) ground plane which is utilized to provide the desired uni-directional radiation pattern while keeping a thin antenna profile. The proposed design offers a 65% reduction in SAR while maintaining a relatively large bandwidth and a compact design.

3. Thermal Effects

Exposure to ionizing electromagnetic radiation, such as from X and Gamma rays, is known to increase the risk of cancer. However, the numerous studies that have examined the potential effects of non-ionizing RF radiation on health from microwave ovens, cell phones, radars, and other wireless transmitting sources confirm that there is no consistent evidence that it could increase the risk of cancer.

The only recognized biological effect of RF energy is hyperthermia, which is defined as an increase in body temperature that occurs when the body generates or absorbs more heat than it dissipates. The ability of microwave ovens to heat water particles in food is one example of the effect of RF energy. While it is confirmed that RF exposure from cell phones does induce heating in the area where they are held (tissues under exposure), it is not sufficiently enough to significantly elevate the user's body temperature.

4. Health Concerns
4.1 Cancer

Some studies report that even radiation from low energy devices could be problematic due to the blood brain barrier effect. According to these studies, exposure to low energy radiation could trigger the opening of the blood brain barrier which allows toxins in the blood stream to penetrate the brain tissues even with low exposure to electromagnetic radiation.

The International Agency for Research on Cancer (IARC), a WHO entity, classifies the use of cell phones as "possibly carcinogenic," based on limited evidence from human and rodents studied samples, and inconsistent results from mechanistic studies. While the National Institute of Environmental Health Sciences (NIEHS) reported that the current scientific evidence that associates cell phone use with any negative health effects is not conclusive, NIEHS did state that more research is needed.

The U.S. Food and Drug Administration (FDA), on the other hand, stated that human epidemiologic studies claiming biological changes linked to RF energy exposure have failed to be replicated, while the U.S. Centers for Disease Control and Prevention (CDC) reported that no scientific evidence conclusively answers whether cell phone use leads to cancer.

The Federal Communications Commission (FCC) confirmed that there is no scientific evidence to establish a link between radiation from wireless devices and cancer. The European Commission Scientific Committee echoes the FCC's report and also concluded that epidemiologic studies do not indicate an elevated risk for other malignant diseases, including childhood cancer.

Other European studies concluded that talking on a mobile phone for extended periods could triple the risk of brain cancer. In contrast, a study published in the British Medical Journal found that there was no proof of increased cancer. However, it is worth noting that the researchers behind the reported study acknowledged that a small to moderate increase in cancer risk must not be ruled out, especially among heavy cell phone users.

4.2 Fertility

Several epidemiological studies have found reductions in sperm quantity, motility, and viability in male subjects using mobile phones for more than a few hours per day. The production of reactive oxygen species (ROS) which can potentially cause damage to cell membranes and DNA was a common finding associated with the reported effects. Another study echoes these finding which found that ejaculated semen from healthy donors exhibited reduced viability and motility, and an elevated ROS levels after one hour of exposure to a cell phone in talk mode.

A more recent study found that exposing ejaculated sperms to WiFi radiation from a laptop for 4 hours led to reduced sperm motility and increased DNA fragmentation as compared with samples exposed to a similar laptop with the WiFi capability turned off. It should be noted that to date, there has been no strong evidence that cell phone radiation affects female fertility.

While most wearables utilize low energy communication schemes, it is the base hubs such devices need to forward, store, and process information (i.e.: cell phone, tablet, etc.) that emit higher electromagnetic energy which could trigger the aforementioned effects.

4.3 Vision and Sleep Disorders

A plethora of studies have found the blue light emitted from electronics such as cell phones, tablets, and laptops reduces the production of the sleep regulating hormone, melatonin. Research findings are clear that people who excessively use their laptops and smartphones are more prone to experience symptoms of insomnia.

Moreover, some studies report that blue light may cause retinal damage which could lead to macular degeneration, a known cause of blindness. Although this damage is attributed to direct exposure which is far greater level of exposure than a user would get from the display of a handheld or a wearable device, researchers have just begun to understand the effects of blue light exposure on vision. Hence, caution should be used especially with emerging head mounted wearables that use projectors and, augmented and virtual reality technologies.

Furthermore, screens on mobile and handheld electronics tend to be smaller than computer displays, which likely forces users to squint and strain their eyes while reading messages or navigating an app. According to The Vision Council of America, more than 70% of Americans are not aware or are in denial that they are prone to digital eye strain, which is a temporary discomfort that follows a few hours of digital device (with a display) use.

4.4 Pain and Discomfort

Some studies correlate joints pain and inflammation to the unnatural rapid movement of hands during the use of handheld device and wearables.

Back and neck pain is also common with increased cell phone use, especially if it is held between the neck and shoulders as the user multitasks. Healthcentral.com which reported that long hours of cell phone use cause the user to arch their bodies in an unnatural posture which can lead to back and pain neck. These abovementioned scenarios are also applicable to wearables that may influence unnatural postures and orientation changes.

Additionally, the findings of some studies suggest that excessive electromagnetic radiation could also stimulate the production of adrenalin and cortisol which may cause headaches, cardiac arrhythmia, high blood pressure, and tremors. However, these effects are generally triggered by high power radiators such as cellular base stations, and are less likely to be caused by low power radiation from wearables.

4.5. Other Risks

One study on fetal development reports that fetuses exposed to electromagnetic radiation from their pregnant mothers' phones can trigger childhood behaviors such as hyperactivity, reduced short term memory, and ADHD.

Although contradicted by other studies, the secretion of cortisol, which is often referred to as the stress hormone, has been shown in one study to be affected by RF exposure. It is assumed that RF radiation may serve as a stressor evident from the elevated cortisol concentrations reported in a number of studies involving animals and humans.

Effects of RF energy localized to the head have been studied repeatedly. Results suggest that RF energy exposure on blood flow in the brain has no hazardous effects. Some studies suggested that RF energy might affect the metabolism of glucose, but follow up studies that examined glucose metabolism inside the brain after cell phone exposure showed inconsistent findings.

Lastly, when it comes to hygiene, the continuous touching of handheld and wearable electronics can foster germs on the device. The greasy residue a user's hand leaves on the device after a day of use can accumulate more disease-causing germs than those found on a toilet seat. For example, one study found that 92% of the 390 smart phones sampled had bacteria on them, 82% of the users' hands had bacteria, and 16% of smart phones and corresponding hands had E. Coli. fecal matter. This would still apply to wearable devices and other gadgets even if it's to a different extent.

4.6. Recommendations

Unlike cell phones, wearables are intended to be in close contact with the user's body, and for far more extended durations. While there is established data on the effects of using cell phones, not enough time has elapsed for research to agree on the health risks and exact impact of wearable devices; hence, it is recommended to use caution and common sense.

More people nowadays are abandoning their landline service and are exclusively relying on their mobile phones. Also, driving laws are regulating the use of mobile devices which has resulted in higher use of Bluetooth ear phones while driving. Therefore, it is advisable to use the built-in hands-free feature (available in most modern cars), wired earphones, or smart home units to minimize low energy radiation effects whenever possible. It is also recommended to avoid positioning any wearable or mobile device in close proximity to the reproductive organs for extended periods of time. These devices should be kept out of pants' pockets. For female users, it is recommended to not place a wearable device within 15 cm of the breast.

The use of mobile phones and cellular-based wearables should be limited to areas with excellent reception. The weaker the reception, the more power the cellular-based IoT or wearable device will have to emit, which means higher electromagnetic energy deposition in the user's body. Obviously, children would be at higher risk from radiation due to their thinner skulls and still developing nervous systems. WHO has stated that the farther away a wireless device is from the user's head, the less harmful it would be.

Lastly, wearing head-mounted wearable devices while asleep and placing IoT and wearable devices on the nightstand next to the head or under the pillow should be completely avoided.

5. Regulations

IoT and wearable devices are still largely unregulated, with no specific laws or regulations governing how this data is collected or used by parties other than the user. In light of the security and privacy concerns associated with these technologies, several bodies have called on the US government, including federal agencies and Congress, to undertake a more active role in coordinating regulation and standards.

Several U.S. agencies including the Department of Commerce, Department of Defense (DoD) and Department of Justice have some form of IoT regulation, but this has led to overlapping responsibilities which created bureaucratic challenges. The need for one inclusionary authority is obvious.

Moreover, experts argue that it is difficult for the industry to develop industry-wide standards due to the fact that the privacy and security of IoT and wearable technology fall upon several actors, including manufacturers, network providers, software developers, and other third parties.

In response to such challenges, the FTC, National Telecommunications & Information Administration (NTIA), Food and Drug Administration (FDA), the National Institute of Standards and Technology (NIST), and representatives from the U.S. Senate and House of Representatives took initiative to address IoT privacy and security concerns by conducting several meeting between 2016 and 2018. Moreover, a public-private sector working group assembled by the NTIA, finalized a guidance document addressing how manufacturers should convey information to consumers concerning security updates for IoT and wearable devices.

Moreover, in 2017, a bipartisan bill aiming at improving the cybersecurity of IoT devices supplied to the U.S. government was introduced by the senate. The bill comprised several provisions, including requirements that supplying vendors guarantee that devices are governed by industry standard protocols, are not based on hard-coded passwords, and do not have any known security vulnerabilities.

The FTC has viewed IoT security as a priority. They provided a set of recommendations for best practices businesses can enforce in order to protect the privacy and security of consumers. In fact, the FTC not only supported giving notice to users about what data is being collected, but also providing them a choice of how their data is to be collected and shared.

Regulation proponents believe that standards applied to every product would help protecting the security of user if the government is to regulate IoT and wearable technology. Others, however, expressed concerns if IoT is to be regulated by the government. These concerns include the elimination of smaller businesses which would compromise market competition and consumer choice, innovation impediment due to bureaucracy, and the lack of government expertise to effectively regulate these technologies.

In addition to calling on the federal government to regulate the IoT and wearable technology, experts have also urged the private sector to engage in self-regulation.

One action was that some of the major industry players such as Google and Sprint were backing the British chip designer ARM's security framework referred to as the Platform Security Architecture (PSA). The objective was to create a common industry framework for every IoT product. According to ARM, about 100 billion devices are already using their designs, and this number is expected to double by 2021. PSA consists of "threat models, security analyses, hardware and firmware architecture specifications, and an open source firmware reference implementation", which, collectively, provide a foundation for security to be consistently integrated into the devices at the hardware and firmware levels.

On the other hand, the European Union (EU) has also taken actions concerning IoT and wearable devices, such as passing the General Data Protection Regulation (GDPR) and the ePrivacy Regulation. Another action was by the European Parliament (EP) where they recommended passing the Privacy Impact Assessments concerning the RFID applications privacy and data protection framework to wearables. The EP also recommended deleting raw data once processed, and immediately informing the user once a data compromise risk is detected.

These regulations exemplify the different approaches pursued by the EU and the United States. The EU uses a holistic approach, providing European citizens with certain privacy rights across all platforms and sectors. The U.S, on the other hand, has an agglomeration of privacy laws specific to different industries. Furthermore, the U.S law typically balances privacy rights and interests against freedom of expression, which is driven by the First Amendment, while the EU firmly asserts that privacy is a fundamental right, and the way personal data is used by third parties should be governed by regulation, controls, and transparency, which requires government supervision.

References

[1] Cellular Phone Towers Center for Health, Environment & Justice FactPack - PUB 129, 2015.

[2] SAR For Cell Phones: What It Means For You, Consumer Guide, Federal Communications Commission, 2016.

[3] Frey, Allan H, "Human auditory system response to modulated electromagnetic energy". Journal of Applied Physiology. 17 (4): 689–692, 1962.

[4] H. Raad, AI Abbosh, HM Al-Rizzo, DG Rucker, Flexible and compact AMC based antenna for telemedicine applications, IEEE Transactions on antennas and propagation 61 (2), 524-531, 2013.

[5] Valentina Hartwig, Giulio Giovannetti, Nicola Vanello, Massimo Lombardi, Luigi Landini, and Silvana Simi, Biological Effects and Safety in Magnetic Resonance Imaging: A Review, Int J Environ Res Public Health, 2009 Jun; 6(6): 1778–1798.

[6] Joseph D. Bowman, Michael A. Kelsh, William T. Kaune, Manual for Measuring Occupational Electric and Magnetic Field Exposires, U.S. Department of Health and Human Services, 1998.

[7] Lennart Hardell, Michael Carlberg, Fredrik Söderqvist, and Kjell Hansson Mild, L Lloyd Morgan, Long-term use of cellular phones and brain tumours: increased risk associated with use for more than 10 years,Occup Environ Med. 2007 Sep; 64(9): 626–632.

[8] Nisarg R Desai, Kavindra K Kesari, and Ashok Agarwa, Pathophysiology of cell phone radiation: oxidative stress and carcinogenesis with focus on male reproductive system, Reprod Biol Endocrinol. 2009; 7: 114.

[9] Saeed Shokri, Aiob Soltani, Mahsa Kazemi, Dariush Sardari, Effects of Wi-Fi (2.45 GHz) Exposure on Apoptosis, Sperm Parameters and Testicular Histomorphometry in Rats: A Time Course Study, Cell J. 2015 Summer; 17(2): 322–331.

[10] http://www.health.harvard.edu/staying-healthy/blue-light-has-a-dark-side

[11] Digital eye strain report, The Vision Council, 2015.

[12] Andrew Goldsworthy, The Biological Effects of Weak Electromagnetic Fields Problems and solutions, March 2012.

[13] Karen N. Peart, Cell phone use in pregnancy may cause behavioral disorders in offspring, Yale News (Article) March 15, 2012.

[14] Biological and Health Effectts of Microwave Radio Frequency Transmissions: A Review of the Research Literature, A Report to the Staff and Directors of the Eugene Water and Electric Board, June 4, 2013.

[15] Nora D. Volkow, Dardo Tomasi, Gene-Jack Wang, Paul Vaska, Joanna S. Fowler, Effects of Cell Phone Radiofrequency Signal Exposure on Brain Glucose Metabolism, JAMA. 2011 Feb 23; 305(8): 808–813.

[16] Jane Kirtley, and Scott Memmel, Rewriting the "Book of the Machine": Regulatory and Liability Issues for the Internet of Things, Minnesota Journal of Law, Science & Technology Volume 19, Issue 2 Article 5, 6-2018.

CHAPTER EIGHT

THE METAVERSE

1. Introduction

The metaverse has been a hot topic of discussion recently, with Facebook and Microsoft both staking claims. We have all heard that that the Metaverse is going to change the way we live, so what exactly is the Metaverse?

The Metaverse is a hypothesized iteration of the Internet, where the real and virtual worlds are fused together through conventional computing, as well as virtual and augmented reality headsets, giving users a persistent online 3-D virtual space to meet, work, shop, and do everything that you can on the Internet and more. The metaverse provides an experience so people can interact with one another in a more immersive fashion, but not physically.

While wearing a VR/AR headset, the user is able to virtually attend a business event or even a concert, as if they were there in person. Aspects of the Metaverse have already been implemented in virtual world platforms of popular games such as Roblox.

The concept of Metaverse is not new. It was first mentioned in the 1992 novel Snow Crash. Later on, a plethora of companies attempted to develop online communities based on the concept, most notably, the game "Second Life", which was released in 2003.

In nowadays somewhat primitive Metaverse, people use avatars to represent themselves, communicate, and build out the virtual community. Users roam a virtual world that mimics aspects of the physical world using technologies such as VR, AR, AI, haptics, and blockchain. Digital currency is used to buy items such as clothes, skins, or weapons and shielding in the case of video games. Users can also virtually explore educational spaces, and travel the world using a VR headset and controllers.

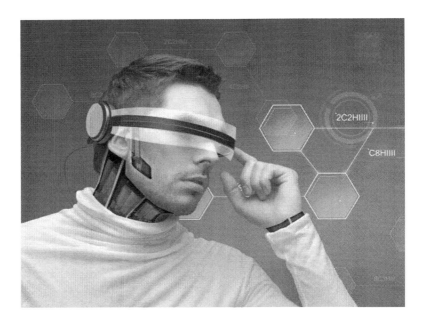

Figure 1: An artistic depiction of a futuristic Metaverse headset

Snow Crash was more of a dystopian view of the future and did not shed light on the Metaverse from a positive angle. The novel's author coined the term Metaverse as a next-generation VR based Internet. One way to attain status in that Metaverse was by growing technical skills, which was expressed by the sophistication of a user's avatar. Another indicator of status was the ability to access certain restricted worlds, a precursor to the paid subscriptions and registration requirements some websites use today.

Another novel, "Ready Player One" by Ernest Cline, which was made into a movie directed by Steven Spielberg in 2011, helped to further popularize the idea of the Metaverse. The dystopian science fiction novel is set in the year 2045, where people escape the disasters plaguing the planet to a virtual world called The Oasis. Users are able to access the Oasis using a VR shield and haptic gloves that let them take a grip of, and touch objects in the digital world.

The growth of the internet has given rise to many services that are leading to the creation of the Metaverse. Conversely, many sees the Metaverse as a successor to the Internet; however, we do not see the Metaverse as the next Internet or posing as its competitor, but rather builds on it and utilizes it to operate. The Internet is something that users "browse", while, to a certain degree, people can "live" in the Metaverse. In a report published in February 2022, Credit Suisse defined the Metaverse as "a more spatially immersive, compelling and frictionless Internet which comprises five essential components: infrastructure, hardware, content, platforms or communities and payment mechanisms.".

The concept of the Metaverse has already spread to the business world and seized the imagination of innovation leaders. Metaverse technology is making its remarkable presence in virtual games, business, education, shopping, fashion, and even the real estate market. The marvelously wide range of potential use cases for Metaverse means it is only a matter of time before the concept becomes mainstream and a part of reality everywhere, and will no longer be limited to gaming and entertainment.

The Metaverse has already cemented itself as an investing theme of the future with giant companies actively investing efforts, capital, and time in developing Metaverse projects. Grand View Research estimates that this industry will grow at a compound annual growth rate of 39.4% through 2030 and reach a market size of $678 billion. Another study by PwC estimates the increase in global economic output through virtual reality and augmented reality to be over $1.5 trillion by 2030.

Meta (formerly Facebook) has spent $10 billion on Metaverse technologies in 2021 alone. The company's efforts include its VR hardware, social VR apps like Horizon in addition to its AR wearables venture. Roblox's proto-metaverse world generated $454 million in the second quarter of 2021 with more than 43 million daily active users. In 2018 and 2019 alone, Epic has made more than $9 billion with Fortnite. It's worth noting that Apple has remained away of the metaverse hype, however, it's evident from the company's tremendous investments in AR hardware that it wants to be part of whatever the future will bring.

The possibilities of the Metaverse will touch all areas of our lives and offer us new ways of living. Virtual venues can be created for events and gatherings in the Metaverse, lands can be sold, houses can be purchased. Immersive travel is possible too. It's crystal clear that the Metaverse is here to stay!

2. Metaverse Characteristics

The virtual space serves to emulate the real world without any physical barriers. Many people have considered various virtual experiences in the past two decadees as the beginning of the Metaverse. So, what are the key characteristics of the Metaverse that make it capable of offering the immersive virtual experiences?

a- **Decentralization**

The Metaverse should never be under the control of a single entity or organization.[28]

b- **Availability**

The Metaverse should always exist regardless of time and place, and allow users to communicate and interact with other users as well as its platforms.

c- **Synchronization**

Users should be able to interact with one another and the digital environment in real time; reacting to their platform and each other just like they would in the physical world.

d- **User Economy**

In the Metaverse virtual worlds, users should be able to create and trade new assets or experiences in addition to enjoying absolute ownership. Creators can use their assets or experiences in any environment and trade them for the desired value using non-fungible tokens (NFTs), cryptocurrency, and the like. Users will also be able to supply goods and services in exchange for value.

---

[28] ***Web 3.0:*** At one time or another, both the decentralized web (Web3) and Metaverse have been referred to as "Web 3.0". The Web3, which embodies the idea of new iteration of the World Wide Web based on blockchain technology, which incorporates concepts such as decentralization and token-based economics is becoming diverse in its offering, ranging from gamers to NFT traders, builders, realtors, etc. While the Web3 aims at providing solutions to the problems of the current internet, the Metaverse is a bigger idea that uses some of Web3's solutions.

e- **Interoperability**

Users should be able to create and use assets, avatars, and experiences in any accessible Metaverse without restrictions. For example, a participant experience may include cross-platform capability, allowing, an asset in a fight game to be used in a different adventure game, or, a skin/clothing item purchased on the Metaverse to be used in other games, concerts, etc.

3. The Elements that Make Up the Metaverse: Hardware and Software Components

In this section, the most important elements in developing the Metaverse are described.

3.1 Hardware

Every emerging generation of technological innovation must be accompanied with an evolution of hardware that brings it to life. Haptic feedback technology is already giving gamers an extra layer of immersion while wallowed in the video games universe. However, today's Metaverse hardware is where cell phones were in the 1980s; bulky, rudimentary, and unwieldy. Fortunately, the motivation is great, the funding is even greater, and enterprises are competing to be the first to get the next phase of digital connectivity to take off.

Hardware includes, but is not limited to, physical user interface hardware such as VR/AR headsets, gaming consoles, and haptic gloves in addition to enterprise hardware such as industrial cameras, projection and tracking systems, scanning sensors, and haptic system/equipment. This also includes computational hardware (microprocessors, microcontrollers, and graphical processing unit) which are used in the calculation, rendering, reconciliation and synchronization of data, machine learning algorithm, etc. It should be noted that such computations could be taking place in the Cloud, Fog, or Mist layer.

3.2 Networking Infrastructure

This includes real-time and wide-band connectivity, decentralized data transmission, data servers, routing centers, and related services.

3.3 Platforms

This includes the creation and operation of immersive three-dimensional environments and worlds wherein individuals and enterprises can explore, socialize, create and participate in a wide variety of experiences (i.e., attending a concert, drive a car, engaging in a business activity, socializing with friends, attending a class, etc.).

3.4 Protocols and Standards

This includes all the protocols, standards, formats and services used to enable the creation, interoperability, compatibility, data management, updating, and all ongoing improvements to the Metaverse.

3.5 Economic/Financial System

The support of digital payment processes, platforms, and operations, financial services, and digital currencies which include cryptocurrencies, such as Bitcoin and Ethereum, and other blockchain technologies.

3.5.1 Blockchain

A blockchain is a series of time-stamped records of data that are linked and secured using cryptography. Each block contains a cryptographic hash of the previous block, and transactions are managed by a cluster of computers not owned by any single entity.

Characteristics of Blockchain:

1- Data contained in a blockchain is resistant to modification.
2- Transactions between two parties are recorded permanently and verifiably by an open, distributed ledger.
3- A blockchain is typically managed by a peer-to-peer network conforming to a protocol for inter-node communication that constantly validates new blocks.
4- Once the data contained in a block is recorded, it cannot be altered retroactively without altering the entirety of all subsequent blocks, which requires consensus from the majority in a given network.

Blockchain nodes are somewhat similar to how smart objects and systems are connected in a network. It treats the data transaction the same way it would treat financial transactions on a Bitcoin network, hence, IoT and wearable systems utilizing blockchain would allow secure, consensus-based messaging between nodes in a network. This will lead to simplifying business processes, improving transparency, providing autonomy, saving costs, and making connected objects such as cars and appliances more reliable and secure.

Here is how blockchain and smart devices can work together:

- **Security:** The ledger used in blockchain cannot be manipulated or tampered which adds another layer of security if implemented in an IoT or wearable system. Moreover, the autonomous security solution blockchain provides makes it a perfect element for IoT and wearable solutions.

- **Decentralization:** The power of a blockchain lies in the fact that there is no single entity controlling the state of transactions. Furthermore, redundancy is forced in the system by ensuring that every node using blockchain maintains a copy of the ledger. Assuming trustless messaging between nodes in a blockchain network, the system must live by consensus.

- **Encryption and Distribution:** The use of encryption and storage distribution in blockchain allow data to be recorded securely in IoT and wearable systems without any human interference which preserves the data integrity allowing it to be trusted by all parties involved in the network.

- **Communication Assurance:** Blockchain allows IoT and wearable devices to securely communicate and exchange transactions with a very high assurance that everything will be processed as per the predefined terms of contract.

- **Cost Saving:** Since no middleman is needed in exchanged and shared blockchain data, significant costs can be saved in the transaction chain. Using smart contracts, blockchain can allow IoT and wearable devices to automate data transactions across various networks.

- **Tracking:** Blockchain can keep unalterable records of the history of an IoT or wearable device. In a network, this property would allow smart devices to autonomously function without the need for a centralized authority. This is just like in cryptocurrencies where direct payment services are provided without the need of any third-party handler.

3.5.2 Non Fungible Tokens (NFTs)

NFTs are a secure type of digital asset based on Blockchain technology. Instead of using currency, an NFT can represent a digital deed or proof of ownership of an asset that can be purchased or sold in the Metaverse (such as piece of art, a song or digital real estate).

Metaverse Properties is regarded as the world's first virtual real estate company. The firm serves as an agent to facilitate virtual properties and lands purchase and rental for a number of Metaverse worlds such as Decentraland, Sandbox, and Upland. Their listings include conference rooms, commercial spaces, art galleries, family homes and virtual cafés!

While the Metaverse has created opportunities for new companies to offer digital products, as in the abovementioned case of Metaverse Properties, established real-world companies are also joining in. For example, Nike has recently acquired RTFKT, a startup company that makes virtual sneakers and digital assets using NFTs, Blockchain, and augmented reality. Nike and Roblox also partnered to form "Nikeland", a digital environment where

Nike fans can play games, socialize, and dress their avatars in virtual clothing.

3.6 Content and Services

This includes the design, development, storage, secure protection and management of digital assets and products as associated with user data and identity.

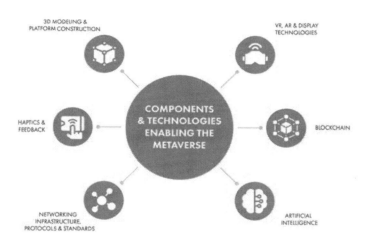

Figure 2: Components and technologies that enable the Metaverse

4. Applications

Gaming and social networking sites are among the most often mentioned use cases of the Metaverse. However, there are several other promising entrants that showcase the real potentials of what the Metaverse can offer in the near future. Let us have a look at the top use cases of the Metaverse across the various industries.

Figure 3: Applications of the Metaverse

4.1 Workplace

The increasing popularity of the Metaverse is giving rise to significant changes to the conventional business processes. Projects that are deployed today will enable:

- **Work from home in a virtual environment:** With the adoption of the Metaverse, it will be unnecessary for employees to be present in the workplace on site.
- **In-person interaction with coworkers in the workplace:** In order to keep up with technological advancements, virtual offices are being established. Further, communication and collaboration in working spaces will be accomplished via VR/AR technology simply, conventional office will become obsolete at some point. This is driven by the growing number of users with diverse industry backgrounds that are interested in practical experience. For example, working on system and equipment maintenance, factory production lines, and bringing all required employees into a single room are examples of VR-first efforts in the manufacturing industry.
- **Holding Remote Meetings:** Currently, such meetings are held over Microsoft Teams or Zoom, but as technology progresses and virtual rooms become more popular, they will eventually be hosted in a virtual room where an avatar will represent each participant. At a later stage, these avatars could be replaced by a realistic 3D model of users that reflects expressions and gestures.

The Metaverse is currently being developed by organizations with insight. According to Illenberger, Virtual Reality showrooms can demonstrate a new truck, automobile, or recreational vehicle instead of traveling to trade shows. A new product in development can benefit from the collaboration of engineers, marketers, salesmen, financial analysts, and creative professionals.

It's worth mentioning that even governments may extend their reach into the Metaverse. For example, while most countries have a relatively static presence on the Internet, Barbados plans to open a diplomatic embassy in the Metaverse realm.

4.2 Fashion

The fashion business, which is considered one of the most progressive sectors of the economy, is continuously introducing new trends geared for the Metaverse technology. On Roblox, one of the most remarkable examples is Gucci, which developed the "Metaverse Design" category. Users can now purchase special outfits from the infamous fashion house while on the gaming platorm. In fact, there was a recent sale of a digital Gucci bag for $4,000, showing that the virtual fashion business is already becoming a considerable source of revenue for the market leaders.

Because of the massive popularity of NFTs, the day will soon arrive when unique virtual collections of the world's most prestigious fashion companies' will be sold for millions of dollars.

4.3 Shopping

Nowadays, many business owners and entrepreneurs are enthusiastic about the potentials of virtual reality, believing that online shopping will soon expand from a two-dimensional experience into an extraordinarily immersive 3D experience. These benefits, including the opportunity to try on, feel and test the goods, as well as the ability to seek online guidance from a seller, will be preserved even though the experience will be free of real-world constraints.

Furthermore, many experts in the fiels believe that one of the most potential uses of the Metaverse technology is virtual shopping, which provides the opportunity to become the owner of the purchased object or service in the actual world. The need to go to shopping malls will no longer be necessary in the near future. Conventional online shopping will be phased out at some point, and virtual shopping at a Metaverse supermarket will become the norm.

In collaboration with a South Korean clothing brand, Ader Error, the Spanish clothing giant, Zara, has launched a new line of collection for people and avatars. Physical and virtual models can now be purchased at network stores anywhere in the world. As mentioned in the previous section, Nike has jumped on the virtual goods wave with "Nikeland", a virtual showroom on the Roblox gaming platform where visitors can dress up their avatars in Nike clothes and shoes, has already been launched by the company.

4.4 Social Networks/Entertainment

Metaverse has the potential to completely transform the worlds of social networks and online entertainment in the years to come. This is simply due to the fact that users in the Metaverse will have a lot more immersive experience in the virtual world than they would have in the current social platforms.

Social platforms will soon grow into fully functioning Metaverses and transform into virtual worlds where individuals will spend their time not only conversing or looking through images but also interacting with avatars or realistic clones of users.

A clear indication of where social networks are headed is the fact that Facebook has already renamed itself "Meta". Also, with the introduction of widely accessible virtual reality capabilities, we can predict a massive development in this area and a significant transition of users away from the regular web and toward the Metaverse.

Games in the Metaverse are centered on the notion of "play to earn", which enables users to win virtual gaming items that they can then sell to get real-world money. Members of the Metaverse may ask their social media friends to join them in playing the games, engage with other members of the Metaverse, and collaborate to enjoy the games as a group. Moreover, players can create goods, sell or purchase them, as well as invest in and compensate other people in the game. As users move through an ecosystem of competing goods, the Metaverse effortlessly mixes games, virtual reality, live-streaming, cryptocurrency, and social media into one seamless experience.

One aspect that has already come to fruition is the fact that real-life musicians is now performing within video games. Minecraft, Fortnite, and Roblox have put up a schedule of concerts by musicians such as Lil Nas X, Royal Blood, 333, IDLES, and Ariana Grande.

Gameplay in a Metaverse setting has the potential to permeate other genres in the future. Examples include the use of virtual reality technology to create a fully-immersive 3D scene for players to walk into at the greatest online casinos in the United States. Like an alternative to choosing a slot game from a list of numerous possibilities, customers may stroll across a virtual gaming floor and insert their digital chips into the machine as they did in real life.

The Metaverse will allow sports organizations to reward their fans by extending events beyond the confines of live games. Fans at digital stadiums in the Metaverse might watch teams and athletes compete in real-time through live streaming. Viewers may even visit the field itself, stroll with the players, join the football cheerleading and watch the game from a variety of vantage points, thanks to multi-view camera technology. In addition, Metaverse enables people to virtually sit in the same room as their friends, even if they are physically thousands of miles apart.

4.5 Tourism

Virtual tourism is one of the most innovative Metaverse applications. Because technology allows you to travel in virtual space, you don't have to go to the destinations that interest you in person to experience them. Experts predict that the development of an immersive digital environment that combines virtual and augmented reality will represent a breakthrough in the field of tourism. A digital area loaded with realistic and immersive content has the potential to become a product that will serve a population of tourism enthusiasts who are unable to physically travel.

We have already witnessed the beginnings of 360-degree virtual tours. Instead of just watching a video tour recorded by a guide, you will be able to experience being there at the destination site. Furthermore, you can go to this location with your family members and friends, which will make your trip much more genuine and enjoyable.

Since its inception, virtual reality has already gained significant traction in the tourism sector. For instance, enabled by the Visualize technology, Thomas Cook introduced the Virtual Reality Holiday 'Try Before You Fly' service, which allowed prospective tourists to experience holiday locations in virtual reality before making the decision to go. After using this service, the client might make a more informed decision about whether or not they wish to go to the destination under consideration. The result was fascinating; for example, after participating in a 5-

minute virtual reality excursion in New York, the number of tours booked in the city more than doubled.

4.6 Healthcare

Using augmented reality to train and educate future medical professionals provides significant benefits in terms of improving their skills and knowledge base and also to lower expenses. In order to facilitate and speed up surgical operations, physicians use medical assistive technologies such as Microsoft Hololens. Additionally, AR headsets are being utilized to view critical real-time patient data such as pulse rate, body temperature, heart rate, and respiratory rate, in addition to pre-operative images from CT, MRI, and 3D scans. In fact, a study shows that VR training for surgeons leads to a performance improvement of more than 230%.

In order to improve vein identification, nurses and doctors are increasingly using augmented reality. This solves the issue of many people having difficulty locating a vein, especially if their skin is very pigmented or if they have small blood vessels. Although the use of visual-driven technologies such as X-rays and CT scans exist, AR will further assist medical professionals in diagnosis and treatment of a variety of conditions.

4.7 Military

Military uses of augmented and virtual reality have also witnessed significant advancements. Night-Vision Goggle (NVG) is a kind of technology that looks to be comparable to tactical Augmented Reality (TAR), but it has far more capabilities. It could show the precise location of a soldier and the locations of allies and opposing forces on a map. The device may be utilized at any time of day or night and is affixed to the helmet in the same way as goggles are. This would eliminate the need for soldiers to look at their GPS position while on the battleground.

Synthetic Training Environment (STE) is an augmented reality system that simulates combat scenarios to offer troops more relistic experience and training by placing them in more physically and mentally demanding battle environments. One of the primary goals of the STE developers is to provide a training alternative that will allow commanders to construct adaptive units that have a greater level of prepation.

4.8 Real Estate

The main benefit here is the ability to provide prospective customers with a realistic and engaging showing experience. Real estate marketing can take advantage of the ability to enable customers to visit the property before making a decision to see it in person. Several multimedia aspects, such as music in the background, narrative, and light-and-sound effects, may also be incorporated into virtual reality tours. These elements provide a sense of realism and excitement in the experience and boost the customer's enthusiasm and confidence before they commit. Through the use of virtual tours and supervised walkthroughs, clients may get an immersive look at the property and its surroundings. This gives prospective buyers and renters a peace of mind regarding a variety of choice criteria, such as traffic conditions or the atmosphere of a location, while simultaneously cutting travel time to zero. Real estate professionals also save time and money by eliminating time-consuming showings and face-to-face interactions with potential buyers and sellers.

In contrast to traditional tours, virtual walkthroughs may be readily customized to appeal to a wide range of client tastes and preferences. Very visual customers will benefit from the lighting, interior design, and zooming features, among other things. The information shown in pop-up windows may be adjusted to provide more detailed information, statistics, and space measurements for data-oriented clients.

4.9 Manufacturing, Training and Occupational Safety

It's shown that virtual and augmented reality technologies make it much less likely for accidents to happen. People learn substantially faster and retain more information when studying in a hands-on setting. Because VR training takes care of everything, manufacturing enterprises do not have to spend extra money or time on training and onboarding new employees. Workers can also understand safety procedures and take part in simulations of dangerous situations and scenarios. These technologies also contribute to developing more effective items in the marketplace. By just wearing a virtual reality goggle, you would be able to evaluate every aspect of the product.

Moreover, virtual reality may be used to develop flooring for manufacturers to use on their production lines. They can immediately determine the most advantageous position for installing the equipment while still maintaining the necessary distance between them. It contributes to the enhancement of safety as well as the proper planning of equipment placements on the job.

The use of VR and AR have helped companies making processes more efficient at lower costs; for instance, BMW utilize these technologies to design vehicle concepts. A virtual tour of a manufacturing plant, for example, was created by Siemens in collaboration with VRdirect. Workers can actively participate in the realistic manufacturing setting by exploring the interactive training environment and interacting with it. Also, Ford employees utilized VR and AR technologies to access their locked-down vehicles when the coronavirus pandamic forced them to stay home.

Lastly, the health and safety of personnel is a major concern for businesses, particularly those that run sophisticated systems and heavy machinery. Employees must be kept informed regularly on industrial security and occupational safety procedures. Here, VR can be used to facilitate training and practicing safe operations with dangerous machineries and equipment. Workers can virtually conduct tests safely before performing it in real life.

4.10 Education

When it comes to highlighting ideas via graphics, traditional teaching techniques will never be as effective as they are with this approach. Regardles of their age, students would always choose to watch a video than read a book. Virtual reality technology has the potential to create fascinating experiences that could never be "experienced" in real life which deem students more motivated in studying if they have access to this technology.

Teachers currently find it incredibly challenging to create a productive and engaging learning atmosphere in the classroom. As virtual reality technology becomes more widely available in schools, the severity of this problem can be significantly reduced since most students will be encouraged to share and speak about their virtual reality experiences. Virtual reality could also assist in the discovery of flaws in the material, as well as providing superior editing capabilities which is becoming more popular. It also eliminates the language barrier which is a substantial obstacle to the educational progress for many students.

4.11 Intimate Relationships

Technology has changed all walks of life, and the COVID-19 pandemic has made that evident. Living and working in the digital world is not only here to stay, but also to change the way we interact, and that includes, intimate and sexual relations, which for sure will have their space in the Metaverse.

If one can attend virtual business meetings, concerts, and venues to interact with others, the possibility would also be for there to be a dedicated space for adult leisure where they can fulfill all kinds of fantasies.

It's well known that the sex industry has historically driven innovative technologies as well as materials and platforms. In the Metaverse, instead of exchanging videos, photographs or having a video call for a sexual encounter, people will be able to have a more immersive and realistic experience through Metaverse enabling technologies such as VR, AR, sensors, actuators, and feedback haptics.

In an infinitely open space where one can interact with people from all over the world, endless possibilities would unfold. This could also be very helpful for people who feel insecure about their appearance; everyone in the Metaverse has the absolute freedom to choose what their avatar will look like. More interestingly, one will also be able to create a 3D realistic facial and body scanning model of themselves. In fact, 3D imaging leverages several supporting technologies, structured light, for example, "reads" a projected light pattern on an object, which is then recreated by scanning cameras that detect distortions caused by differences in distance.

Also, in order not to lose that physical bond and kep the relationship lively, individuals in a long-distance relationship could also have immersive sexual encounters in the Metaverse.

The Author anticipates that 3-D modeling and scanning companies would even offer celebrities to scan their faces and bodies along with gestures and movements which will then be available for purchase as a package to the Metaverse citizens who might be interested a specific fantasy.

Human-Machine Interface (HMI) is already undergoing an incredible transformation. Just recently, Meta's Reality Labs division showcased a prototype of a haptic glove with ridged pads that allow the user to "feel" surface textures. Optics and displays are also taking an innovative leap. In fact, transparent lenses of combiner optics can now overlay the real world with virtual images, which will make AR glasses more practical. See-through waveguides on high-refractive-index glass can also be enabled by surface relief waveguide technology, which will support 3D sensing, and automotive heads-up and VR displays.

Another marvelous technique called Time of Flight (ToF) is utilized in this area. It works by emitting light pulses into the virtual environment which are then reflected and detected by a sensor, enabling moving objects to be identified and integrated into the scene.

To conclude, such technologies along with the fast-paced world of haptics and "real-feel" sex toys and equipment, virtual sex could reach its greatest potential.

5. Concerns and Technical Challenges

It is clear that the Metaverse is a promising technology that can provide fully immersive experience, elements of fantasy, and new degrees of freedom. However, it is still considered controversial since it will also open up opportunities for misconduct and crime. Furthermore, the industry lacks the capacity to conduct a comprehensive study of the potential risks. The section below presents the main concerns and challenges faced by the Metaverse stakeholders.

5.1 Concerns

5.1.1 Privacy

As mentioned in the previous chapter, information privacy is one of the biggest concerns for Wearable Technology and IoT. The Metaverse is no exception since it is enabled by these technologies. According to a study, 55% of U.S. adults said the tracking and misuse of their personal/biometric data is a major concern. Meta (formerly Facebook) is planning to employ targeted advertising within their Metaverse platform, raising further concerns related to the spread of misinformation and compromising personal privacy.

5.1.2. Addiction

User addiction is another big concern. We showed in the previous chapter that Internet, social media, and videogaming addiction can have negative mental and physical consequences over a prolonged time, such as depression, anxiety, and various other disorders related to having an inactive and seat-bound lifestyle (i.e.: an increased risk for obesity and cardiovascular diseases).

A marketing professor, Andreas Kaplan, who studied the user experience of the game "Second Life" believes that the Metaverse may have an overall negative societal impact due to their considerably addictive potential.

Psychologists and social experts are also concerned that the Metaverse could be used as an 'escape' from reality similarly to what is happening currently with existing Internet technologies. Moreover, being inside a virtual space, a user may perceive time and space differently. Similar to gaming, full immersion in the Metaverse can motivate users to spend much more time than they had planned initially, raising the question of how the user will perceive reality while experiencing a distorted version of time inside the Metaverse. As for spatial perception, the issue here is that the Metaverse space is theoretically infinite, which means new users will find it challenging to adapt to a massive volume and diversity of information.

5.1.3. User Safety

Virtual crime and social media misconduct such as cyberbullying and sexual harassment are significant challenges facing the digital social space currently, and will for sure be prevalent in the Metaverse.

In 2022, investigations by The Washington Post and BBC News found under-age users engaging in adult activities in applications such as VRChat and Horizon Worlds despite an age restriction (13 years or older). Other major concerns include the potential presence of child predators hiding behind friendly looking avatars in the Metaverse platforms, along with other potentials for child depression and loneliness. In fact, according to the same study reported in the previous section, cyberbullying and online abuse in the Metaverse was the second-biggest worry, with 44% of participants indicating it was a major concern.

5.1.4. Social Issues

In 2022, The Guardian newspaper criticized the utopian mentality of tech companies who claim that the Metaverse could be a solution to worker exploitation, prejudice, and discrimination problems. The newspaper reported that they would be more supportive towards the development of the Metaverse if it was not dominated by "companies and disaster capitalists trying to figure out a way to make more money as the real world's resources are dwindling".

5.1.5. Identity and Reputation Concerns

Theoretically, any person or even a bot can easily impersonate other users in the Metaverse. Hence, reputation should be a crucial element of authentication and certificate of confidence concerning any object or user in the virtual space.

5.1.6. Security

With the full emergence of the Metaverse it will be crucial to dramatically raise the cybersecurity measures. There will be a need to create new approaches and techniques to protect personal data, privacy, and digital assets. It is anticipated that users in the Metaverse may be required to provide much more personal data (even biometric) than they would do currently in order to enforce more accurate identification and enhanced security.

5.1.7. Financial System

Just like in the real world, individuals in the Metaverse could engage in financial transactions, and hence major financial companies are focusing on integrating their services in the virtual space. It is already established that cryptocurrencies will ensure fast and secure exchange in the Metaverse which will carry out its own class of the virtual market. This will create the need for additional unique transaction verification system to safeguard the financial security of users.

5.1.8. Regulation and Legal Issues

Establishing the regulations and laws that will govern the virtual world is of paramount importance. The Metaverse is extremely vulnerable when it comes to the rights of its participants, since it is not legally regulated as of today. According to one study, 36% of participants indicated that governmental regulations would be key when considering whether they want to be part of the Metaverse, as lawmakers foresee initial regulatory steps for the new technology.

5.2. Technical Challenges

5.2.1. Hardware

In this context, hardware includes the physical technologies and devices needed to use, interact with, or develop the Metaverse. This includes, but not limited to, the user-interface hardware (i.e.: VR/AR headsets, handheld controllers, and haptic gloves and body suits), as well as the enterprise hardware used to operate or create the virtual environments such as cameras, projectors, optical equipment, tracking devices, and scanning systems.

Although consumer electronic hardware is constantly improving and getting equipped with more powerful processing capability, more superior sensors and haptics, higher resolution screens, sharper cameras, and longer battery life, industry experts still believe that hardware constraints are what's holding the Metaverse back from realizing its full potential.

Enabling a 3-D environment in the Metaverse is highly dependent on virtual or augmented reality technologies. Such technologies already exist today, but a great deal of work is needed to get them to be mainstream. To that end, companies are actively working towards such improvements. For example, Meta has significantly improved its Oculus VR headsets in recent years by eliminating the need to be constantly plugged to a gaming box or a computer. However, even the most expensive models are still not swift enough with their head tracking, and in desperate need for more powerful graphic processor units (GPUs) to solve this issue.

In addition to enhancing and creating more powerful GPUs to be embedded into VR/AR headsets, eye-tracking technology has a great potential when it comes to reducing the load placed on the hardware. Furthermore, VR/AR headsets are still too bulky and heavy to wear for extended periods of time, meaning your immersive sessions in the Metaverse would have to be kept brief. To conclude, manufacturers need to find ways to balance the needs for powerful hardware and a lightweight product.

Beyond the processing requirements to create hardware that lets us live in a world that transitions seamlessly between digital reality and real-life lies a big step forward in display technology.

If you've used VR before, you may have noticed that gaps appear in the image. This is because the pixels that make up the image are not close enough together. A smartphone's display is sufficiently good for looking at from an arm's length away, but when the display is an inch or two from your eyes, it will require at least 60 pixels per degree of field of view. Moreover, most of today's display technologies are not bright enough to provide a sharp AR image.

5.2.2. Networking

The Metaverse products and services face three main networking constraints: bandwidth, latency, and reliability.

5.2.2.1. Bandwidth

Bandwidth is commonly thought of as 'speed', but it's actually refers to the maximum amount of data a connection can handle at any moment. Speed refers to the maximum rate data can be transmitted, typically measured as megabits per second (Mbps) and gigabits per second (Gbps).

Obviously, the requirements for the Metaverse are much higher than most nowadays Internet applications and games, and is expected to drastically increase as the complexity of virtual processes grows.

Currently, some platforms benefit from the fact that a number of previously-made digital assets are widely repurposed and slightly modified/customized. For example, Roblox is mostly streaming data on how to modify previously-downloaded objects. However, the virtual platform will eventually require a near-infinite number of transformations and creations. Many online gamers already struggle with bandwidth and network congestion in games that require only positional and input information. The Metaverse will further intensify these needs. The good news is that broadband connectivity and bandwidth is continuously improving worldwide. Machine learning algorithms fueled by the ever-improving computational power can also help substitute for constrained data transmission by predicting what output should be triggered while waiting for the actual input data from the user.

5.2.2.2. Latency

Latency, one of networks' biggest challenges, refers to the time it takes for data to travel from one node to another (round trip). In most modern applications, it does not matter if it takes 100 ms or 200 ms or even three-second delay from the time a WhatsApp message is sent and a read receipt is received. Likewise, it does not matter if it takes up to 500 ms after a user clicks YouTube's skip button until the next video starts. On the other hand, when watching a subscription streaming service like Netflix or Hulu, it's more important that the show plays continuously without buffering than played right away. To that end, streaming services delay the start of a video stream on purpose so that your device can download ahead of the very instant you're watching. That way, if the network encounters a congestion or hiccup for a second or two, you'll never notice. Unfortunately, latency has bigger impact when it comes to Metaverse environments where fluidity and real-time feel are of paramount importance.

5.2.2.3. Reliability

By reliability, we mean the consistency with which the connectivity is available. Our ability to shift to a virtual environment where business, education, and entertainment take place is highly dependent on a reliable quality of service. This covers both overall uptime (the time when a network is up and running), as well as the consistency of other aspects of the network such as download/upload speed and latency. That said, subscription services such as Netflix streams in 1080p or even 4K quite well most of the time. However, it should be noted that such services leverage reliability solutions (such as the buffering example we mentioned in the previous section) that will not work well for gaming or Metaverse applications.

To conclude, obstacles and challenges faced by the Metaverse developers are likely to prove solvable in time. However, they all require high ethical standards and original engineering solutions from the architects.

Conclusion

Humanity is on the verge of entering a new age. It is expected that virtual reality will become more and more incorporated into our daily lives. Currently, the notion of the Metaverse virtual worlds has expanded well beyond the field of immersive gaming, and it is increasingly finding its way into fields such as entertainment, communications, business, education, and other creative industries.

Meanwhile, building a virtual world presents a number of concerns and obstacles that have yet to be answered, even as it reveals a great deal of potential. There must be no attempt to replace the actual world but rather to make it better in every manner conceivable via the use of the Metaverse. The developers have the primary responsibility in this regard.

References

[1] Matthew Ball, Framework for the Metaverse: The Metaverse Primer, Jun 29, 2021.

[2] Anders Bruun and Martin Lynge Stentoft. Lifelogging in the wild: Participant experiences of using lifelogging as a research tool. In INTERACT, 2019.

[3] William Burns III. Everything you know about the metaverse is wrong?, Mar 2018.

[4] Kyle Chayka. Facebook wants us to live in the metaverse, Aug 2021.

[5] Nvidia omniverse™ platform, Aug 2021

[6] F. V. Langen. Concept for a virtual learning factory. 2017.

[7] Aaron Bush. Into the void: Where crypto meets the metaverse, Jan 2021.

[8] S. Viljoen. The promise and limits of lawfulness: Inequality, law, and the techlash. International Political Economy: Globalization eJournal, 2020.

[9] Ying Jiang, Congyi Zhang, Hongbo Fu, Alberto Cannavo, Fabrizio ` Lamberti, Henry Y K Lau, and Wenping Wang. HandPainter - 3D Sketching in VR with Hand-Based Physical Proxy. Association for Computing Machinery, New York, NY, USA, 2021.

[10] Michael Nebeling, Katy Lewis, Yu-Cheng Chang, Lihan Zhu, Michelle Chung, Piaoyang Wang, and Janet Nebeling. XRDirector: A Role-Based Collaborative Immersive Authoring System, page 1–12. Association for Computing Machinery, New York, NY, USA, 2020.

[11] Balasaravanan Thoravi Kumaravel, Cuong Nguyen, Stephen DiVerdi, and Bjorn Hartmann. ¨ TutoriVR: A Video-Based Tutorial System for Design Applications in Virtual Reality, page 1–12. Association for Computing Machinery, New York, NY, USA, 2019.

[12] Haihan Duan, Jiaye Li, Sizheng Fan, Zhonghao Lin, Xiao Wu, and Wei Cai. Metaverse for social good: A university campus prototype. ACM Multimedia 2021, abs/2108.08985, 2021.

[13] John Zoshak and Kristin Dew. Beyond Kant and Bentham: How Ethical Theories Are Being Used in Artificial Moral Agents. Association for Computing Machinery, New York, NY, USA, 2021.

[14] Semen Frish, Maksym Druchok, and Hlib Shchur. Molecular mr multiplayer: A cross-platform collaborative interactive game for scientists. In 26th ACM Symposium on Virtual Reality Software and Technology, VRST '20, New York, NY, USA, 2020. Association for Computing Machinery.

[15] Lik-Hang Lee, Tristan Braud, Farshid Hassani Bijarbooneh, and Pan Hui. Ubipoint: towards non-intrusive mid-air interaction for hardware constrained smart glasses. Proceedings of the 11th ACM Multimedia Systems Conference, 2020.

[16] Aakar Gupta and Ravin Balakrishnan. Dualkey: Miniature screen text entry via finger identification. Proceedings of the 2016 CHI Conference on Human Factors in Computing Systems, 2016.

[17] Yizheng Gu, Chun Yu, Zhipeng Li, Weiqi Li, Shuchang Xu, Xiaoying Wei, and Yuanchun Shi. Accurate and low-latency sensing of touch contact on any surface with finger-worn imu sensor. Proceedings of the 32nd Annual ACM Symposium on User Interface Software and Technology, 2019.

[18] J. Gong, Y. Zhang, X. Zhou, and X. D. Yang. Pyro: Thumb-tip gesture recognition using pyroelectric infrared sensing proc of the 30th annual acm symp. on user interface soft. and tech. (uist '17). pages 553–563, 2017.

[19] Farshid Salemi Parizi, Eric Whitmire, and Shwetak N. Patel. Auraring: Precise electromagnetic finger tracking. Proc. ACM Interact. Mob. Wearable Ubiquitous Technol., 3:150:1–150:28, 2019.

[20] Eduard Fosch-Villaronga and Adam Poulsen. Sex robots in care: Setting the stage for a discussion on the potential use of sexual robot technologies for persons with disabilities. In Companion of the 2021

ACM/IEEE International Conference on Human-Robot Interaction, HRI '21 Companion, page 1–9, New York, NY, USA, 2021. Association for Computing Machinery.

[21] Nina J. Rothstein, Dalton H. Connolly, Ewart J. de Visser, and Elizabeth Phillips. Perceptions of infidelity with sex robots. In Proceedings of the 2021 ACM/IEEE International Conference on Human-Robot Interaction, HRI '21, page 129–139, New York, NY, USA, 2021. Association for Computing Machinery.

[22] Giovanni Maria Troiano, Matthew Wood, and Casper Harteveld. "and this, kids, is how i met your mother": Consumerist, mundane, and uncanny futures with sex robots. In Proceedings of the 2020 CHI Conference on Human Factors in Computing Systems, CHI '20, page 1–17, New York, NY, USA, 2020. Association for Computing Machinery.

[23] Anna Zamansky. Dog-drone interactions: Towards an aci perspective. In Proceedings of the Third International Conference on AnimalComputer Interaction, ACI '16, New York, NY, USA, 2016. Association for Computing Machinery.

[24] http://www.trade.gov/markets/smartcities.pdf

[25] Nick Hunn, The Market for Smart Wearable Technology A Consumer Centric Approach, WiFore Consulting, February 2015.

Published by Ohio Publishing and Academic Services, USA
Copyright © 2022 Ohio Publishing and Academic Services
www.ohiopas.org
info@ohiopas.org

All rights reserved. No part of this book may be reproduced, copied (electronically or mechanically), or videotaped without a written permission of the publisher.

ISBN: 978-1-7372334-8-0

Disclaimer
The Publisher and the Author hold no liability for incidental or consequential injuries or damages caused by the information contained in this publication.

Email: info@ohiopas.org
www.ohiopas.org

Made in the USA
Monee, IL
22 December 2023